Detecting the Bomb

Also by New Academia Publishing

Foreign Affairs / History

THE SOVIETIZATION OF EASTERN EUROPE: New Perspectives on the Postwar Period, edited by Balázs Apor, Péter Apor, and E. A. Rees, eds.

WITNESS TO A CHANGING WORLD, by David D. Newsom

AN ARCHITECT OF DEMOCRACY: Building a Mosaic of peace, by James Robert Huntley

ECHOES OF A DISTANT CLARION: Recollections of a Diplomat and Soldier, by John G. Kormann

NINE LIVES: A Foreign Service Odyssey, by Allen C. Hansen

BUSHELS AND BALES: A Food Soldier in the Cold War, by Howard L. Steele

ARIAS, CABALETTAS, AND FOREIGN AFFAIRS: A Public Diplomat's Quasi-Musical Memoir, by Hans N. Tuch

To read an excerpt, visit: www.newacademia.com

Detecting the Bomb

The Role of Seismology in the Cold War

Carl Romney

NAP   NEW ACADEMIA
      PUBLISHING

Washington, DC

New Academia Publishing, 2009

Printed in the United States of America

Library of Congress Control Number: 2008938047
ISBN 978-0-9818654-3-0 hardcover (alk. paper)

New Academia Publishing
P.O. Box 27420
Washington, DC 20038-7420
info@newacademia.com - www.newacademia.com

Contents

Acknowledgments

I wish to acknowledge my great appreciation to Jack H. Hamilton, who established the Hamilton Visiting Scholars Program and who, incidentally, was a colleague in developing bomb-detection capabilities for the United States in the early years and for some years to follow. My thanks also to Professor Eugene Herrin, who directs the Program, and to a number of students and staff members of SMU who assisted me in many ways, including making my visit a stimulating experience.

Several colleagues read drafts of the early chapters of this book and provided comments to help me set the technical level. These included Jan Gaudaen, Dennis Gormley, Ann Kerr, Carolyn Lock, Keith McLaughlin, Herbert Robertson, and Xiaoping Yang. Billy Brooks, William Braukman, and Jack Hamilton, all participants in the early days of U.S. nuclear test detection efforts, did the same and also gave personal recollections and insights into the work. Seismologists Robert Blandford and Donald Springer, each of whom has devoted a full career to seismic effects of nuclear testing, have read my conclusions and given valuable suggestions. Historians Michael Goodman and Frank Panton, affiliated with the University of Nottingham, and Kai-Henrik Barth of Georgetown University—each broadly knowledgeable of my subject matter—have seen early drafts of my entire manuscript and given helpful comments. Russian colleagues Dzhamil Sultanov and Alexei Vasiliev, have each sent me information on early Soviet nuclear test detection efforts, shedding light on people and work formerly obscured by security. Carole Sargent gave me valuable assistance in preparing a book proposal and in identifying potentially interested publishers. Mark Bernstein critically reviewed the entire manuscript, and made numerous corrections and editorial suggestions. Finally, this work would have been impossible without the essential help of Holly Lanigan at all times during the preparation of the manuscript. My heartfelt thanks to all of these friends and colleagues.

Acronyms

ACDA: U.S. Arms Control and Disarmament Agency.

AEC: U.S. Atomic Energy Commission.

AEDS: U.S. Atomic Energy Detection System.

AFMSW: Air Force Material Command, Special Weapons Group-1, the initial designator of the U.S. long range detection Research and Development organization prior to 1 July 1948.

AFOAT-1: Air Force Office of Atomic Energy-1, the designator of the U.S. long range detection organization from 1 July 1948 to 7 July 1959.

AFTAC: Air Force Technical Applications Center, the designator of the U.S. long range detection organization from 7 July 1959 to the present time.

ARPA (or DARPA): Advanced Research Projects Agency of the U.S. Department of Defense.

B&H: Beers and Heroy, consulting geologists, geophysicists, and engineers.

DOD: Department of Defense.

ENDC: Eighteen Nation Disarmament Conference of the U.N.

GSE: Group of Scientific Experts, an adjunct to the conference on disarmament.

LAC: Laramie Analysis Center.

LRD: Long Range Detection.

LRSM: Long Range Seismic Measurements, the U.S. research program to measure seismic waves from nuclear explosions.

NTS: Nevada Test Site.

PSAC: President's Science Advisory Committee.

SMS: Special Monitoring Service of the Ministry of Defense of the USSR —organized to monitor foreign nuclear testing.

USC&GS: U.S. Coast and Geodetic Survey.

WWSSN: Worldwide Standard Seismic Network, international network equipped in the 1960s with U.S.-made seismographs for obtaining data to support research.

Preface

This book is a perspective on early attempts to monitor and then to ban atomic weapons tests, as well as on scientific factors leading, instead, to a treaty in 1963 banning tests in all environments except deep underground. The early part of this interval was one in which the great alliance of powers that won World War II split apart and descended into what became known as the Cold War. The bomb that had ultimately ended the war in the Pacific was a pivotal element in the disagreement between East and West. When the original developers of the bomb—the U.S., the U.K., and Canada—offered in 1945 to give up their monopoly on atomic energy and destroy all weapons in exchange for controls that would ensure that all further uses would be for peaceful purposes, they were rebuffed by the Soviet Union. Instead, the Soviet Union retreated behind an "Iron Curtain" of secrecy following the war. It became imperative for the West to discover if and when the Soviets developed nuclear weapons, necessitating the development of scientific and technical methods to detect nuclear tests at distances of thousands of kilometers.

This perspective on the science and the times is a personal perspective. I was fortunate to be a graduate student in seismology even as the first scientific papers were being published on seismic observations of atomic explosions. Reprints of these and other significant articles are still in my collection; thanks to the small population of seismologists of that day, even lowly graduate students could receive reprints from such eminent seismologists as Beno Gutenberg and Charles Richter (Caltech), Perry Byerly (University of California), Dean Carder (U.S. Coast and Geodetic Survey), Keith Bullen (University of Sydney) and L. Don Leet (Harvard University).

I became involved professionally in developing the science and art of bomb detection in 1949 when the work was highly classified, and through this involvement gained a window into research and development focused on bomb detection years before it was mentioned substantively in the unclassified scientific literature. It was a time when there was virtually everything still to learn about bomb detection. Later, during the years after much of the research became public, I maintained a number of notebooks and collections of unclassified papers that have been a fruitful resource for this work. More recently, I have been able to obtain or view a number of declassified documents pertaining to my subject. I have used and given a number of references to books and journal articles, in most cases to identify either a source of more information or a key stepping-stone in our thinking about the various subjects to be described here. A few references, mostly post-1960, have helped me to remember and describe the international political background of the era in which these developments took place. However, the main themes and events of this work are those in which I was personally involved. Thus, it remains as a perspective from my own point of view, and I make no claim for completeness.

One focus for this book is the early development of the seismic component of the U.S. Atomic Energy Detection System (AEDS), our then-secret system for detecting and determining the characteristics of foreign atomic explosions. This has guided my selection of events and research results to include here, especially in the years before 1958. While working with the consulting firm of Beers and Heroy, I was instrumental in the design of an experimental seismic system for long range detection. In 1951, together with our sponsoring agency, now known as the Air Force Technical Applications Center (AFTAC), we validated the system during the GREEN-HOUSE nuclear test operations. AFTAC then rapidly deployed seismic stations in countries surrounding the U.S.S.R., and in August 1953, the still-evolving seismic system detected and accurately located the fourth Soviet nuclear test. However, this book is not intended as a history of the seismic system of the AEDS, but rather a disclosure of seismological research and development that impacted the AEDS, as well as early test ban treaty negotiations.

I have included, very selectively, some of the political circumstances and events within the U.S. and internationally that led to the creation of, and influenced the early years of, the AEDS and its creator and operating agency, AFTAC. In selecting which of the earliest events to include here, I have been guided by the "folk lore" I absorbed from those directly involved in the two years before my time, especially the late Doyle L. Northrup, Technical Director of AFTAC during most of the years of my close association. I gratefully acknowledge major help from three other sources in placing accurate dates, names, and sometimes context on material of this sort. These sources are: AFTAC (1997), Ziegler and Jacobson

(1995) and Welch (1996, 1997); I highly recommend these references to those interested in the history of the AEDS.

A second focus is on the influence of atomic explosions on seismology in general. Sensitive detection and reliable discrimination between explosions and earthquakes required addressing fundamental aspects of classical seismology -- the nature of microseismic "noise" in the earth, the mechanisms of earthquakes and how they excite seismic waves, the effect on seismic waves of propagation through the heterogeneous earth, to mention a few. The atomic explosions themselves became an invaluable part of many investigations of these fundamental problems. For the first time sources of seismic waves with precisely known locations, times and energies—and large enough to be recorded at great distances—became available. These led to fundamental discoveries in the science, and to improved knowledge of the structure of the earth itself.

A third, and perhaps the most important, focus is on consequences of the international political thrust in the 1950's toward an agreement banning—or limiting, as it turned out in 1963—nuclear testing. Public concern about radioactive fallout was an important factor in this, and has sometimes been cited as the major one, or even the only one. But I am persuaded that a larger factor in the thinking of the world's leaders was the increasing size of both weapons and arsenals, and genuine fear that they might be used through miscalculation under the tensions engendered by the East-West rivalry. A nuclear test ban seemed a reasonable first step toward abating an unbridled arms race. However, the stakes were high, and neither side wanted to give the other an advantage in weapons development. In a world where information was tightly controlled from a continental-sized unfriendly region, confidence within the U.S. that adherence to a test ban agreement could be verified would depend on confidence in scientific monitoring capabilities.

At President Eisenhower's direction, an attempt to pave the way for test ban treaty negotiations brought American scientists involved in test detection into intensive contact with their Soviet counterparts. The resulting Conference of Experts in 1958 broke negotiating precedents in a number of ways, and established scientific verification as a cornerstone of test ban treaties and of a number of other treaties to follow. These meetings, held in Geneva, Switzerland, established that substantial agreement existed on methods and capabilities for detecting atmospheric tests, where several methods apply in both overlapping and complementary ways. But significant disagreement between scientists of East and West was revealed on detecting deep underground nuclear tests. For these, seismology is the only means for detection at long ranges; obtaining proof of *nuclear* origin requires the use of other techniques, and can be very difficult. Resolution of these disagreements was not possible with data from the single underground nuclear explosion that had been detonated up to that time.

New data from several additional U.S. underground explosions, acquired only days before treaty negotiations began in 1958, kept the seismological issues prominent. New problems were revealed by the new data, and once again scientists of East and West were unable to resolve differences in their judgments on nuclear test monitoring capabilities. These differences, in the context of profound political differences between East and West, became obstacles that strongly influenced the course of nuclear test ban treaty negotiations for many years to come.

When deficiencies in seismic knowledge were seen by high U.S. national political leaders to create obstacles to a much desired treaty, they greatly increased funding for the science. What followed was an interval in which classical seismology developed in scope from an obscure science, centered mainly in a few universities, into a flourishing and highly sophisticated science also involving industrial, governmental, and federally funded laboratories. Seismographic data recorded photographically (and even on smoked paper) and analyzed visually were rapidly being replaced by magnetic tape recordings, and modern digital computers became available for analyzing the data. At the beginning of this period attendees of the Seismological Society of America's annual meetings were easily accommodated in a small lecture room and were on a first name basis; within a few years, seismologists were multiplying at a rate that made both impossible.

U.S. scientists conducted their research against the backdrop of treaty negotiations, and as early research results emerged—not all of them good news for treaty advocates—U.S. policy-makers and negotiators were forced to struggle at times to adjust to the changing technical picture. To the USSR, however, the treaty was predominantly a political matter, and the US stress on science was vehemently criticized as merely a pretext to undermine the negotiations. To finesse the underlying seismological issues, the negotiators compromised in 1963 on a ban of all but underground tests.

What follows is loosely historical in structure. However, it has seemed logical to me to separate technical and political occurrences in some cases and follow one thread to some milestone before taking up another set of events in the same time frame. This can lead to viewing one set of events from two perspectives: the advent of thermonuclear tests impacted seismology beneficially, but its attendant radioactive fallout strongly impacted the political world in quite a different way.

There is no precise moment of genesis for this book. A number of colleagues have urged that I do something of this sort, starting at least twenty years ago. Somewhat later Mikhail A. Sadovsky, a leader of the Soviet Union's test detection research since its inception, suggested that we write a joint article on subjects discussed and issues argued during and after the Geneva Conference of Experts of 1958, but that was not to be because of his failing health. The main impetus that actually led to putting words on paper came at the end of 1996, when I was asked to consider initiating the Hamilton Visiting Scholars Program at Southern Methodist University. After accepting early in 1997, I began sorting through old files, reprints, notebooks and so on, and collecting and organizing this and other material as the basis for a series of lectures/seminars. Seven lectures were prepared, and delivered at SMU from hand-written notes (my keyboard skills are distinctly minor) during October and November of 1997. Illustrations were in the form of copies of hand-drafted figures from old reports for the most part—a practice also followed here to help illustrate the technology and informal style of that time, well before the advent of computer-aided graphics. What follows has its origins in those lectures.

ONE

From *Trinity* to *Joe-1*

The Beginning

Seismic waves from a nuclear test were first recorded at Journada del Muerto Valley (Journey of Death Valley), northwest of Alamogorda, NM. The test took place in the early morning hours of Monday, 16 July 1945, a 21-kiloton (kt) nuclear device detonated on a 100 foot tower. The event, of course, was the *Trinity* explosion,

Trinity remained a secret event while preparations went on to use the atomic bomb against Japan in the hope of ending the war in the Pacific. I first learned of it on August 7, shortly after the bombing of Hiroshima. At the time, as a midshipman in the U.S. Navy stationed at the University of Notre Dame, I was on a training cruise in a small Navy patrol craft on Lake Michigan. For most of us, about to become officers in the U.S. Navy, it was a surprise, and a relief. Although we were all volunteers, and dedicated to serving as needed, most of us had plans beyond serving in the Navy—in my case, to attend graduate school. I had no inkling how much of my future work would revolve around events like *Trinity*.

But, back to the *Trinity* test, where Professor L. Don Leet of Harvard University had installed seismographs at several distances from the intended shot point. Leet specialized in measuring ground motions near quarries and sites of other large industrial blasts, and he had designed his own seismographs for measuring the relatively large motions near the shot point. His measurements were used to validate or dispute claims of damage to nearby man-made structures. He had been asked to make similar measurements on *Trinity*.

As Leet (1946) would later write, "On July 16, 1945, Lamb's conditions were met experimentally for the first time." He was referring to Horace Lamb's (1904) classical paper in which the eminent English mathematical physicist had solved the elastic wave equations to predict the seismic waves caused by "tapping" the surface of a flat, homogeneous solid of great extent in all directions. Lamb predicted that three types of waves would be generated, each traveling with its own characteristic speed so that three simple pulses might be detected at some distance from the source. Wave motion for each would be confined to the vertical and radial directions. Seismologists commonly refer to these as the "P-wave," the "S-wave," and the "Rayleigh-wave" (see Appendix A and Glossary for further description).

The waves Leet recorded by seismographs placed 8.2 km north of the shot didn't conform at all to the predicted sequence of simple P, S, and Rayleigh pulses all within a few seconds, if Lamb's conditions had actually been met. Instead, there were continuing oscillations of the ground following the arrival of the P-wave, growing to a maximum after about 11 seconds, and only slowly decreasing after that. Furthermore, there were large waves on the transverse component (motion perpendicular to a line between the source and station). Had the earth conformed to the overly simplified model used by Lamb, there should be no transverse motion at any time.

Seismology's "New Wave"

Leet analyzed the details of the motions of the earth's surface as these late waves passed his seismograph, and concluded that some of the motion was not predicted by standard elasticity theory. In particular, he pointed out two short intervals in which the motion did not conform to well-known wave types. Hoping that he had discovered fundamentally new types of waves, he named them the "coupled wave" and the "hydrodynamic wave."

Referring to an earlier paper that he had written, Leet noted rather plaintively that although the coupled wave was "first reported in 1939 [by him, of course], no attention has been paid to it since." The observed motion was "along the diagonal of a rectangular frame," with motion "Right Up Push- Left Down Pull." (Here he means transverse motion to the right simultaneously with vertical motion and radial motion away from the source, followed by the reverse.)

Concerning his hydrodynamic wave Leet declared, "This is a new wave to seismology." The earth particle motion describes a prograde ellipse in the vertical-radial plane, with the major axis horizontal. (Picture the motion of a floating cork on a smooth pond as a train of water waves from a tossed pebble passes by.) This is in contrast to the Rayleigh wave in which an earth particle motion describes retrograde elliptical motion in the vertical-radial plane, with major axis vertical.

Both anomalous in-phase and elliptical motions have since been seen from other explosions (Howell, 1949). They are not well understood. However, Leet's hopeful claim that he had discovered new wave types outside of elastic theory has not been accepted. Instead, we tend to ascribe such motion to scattering, reflections and refractions from tilted layers in the underlying rock, or the presence of higher mode surface waves, etc., rather than to fundamentally new wave types. And, while the explosion itself met Lamb's conditions of a "tap" to the surface, the earth structure underlying Journada Del Muerto Valley is far from homogeneous, as is true of almost any place in the real world.

However, Leet's recordings—no matter how we interpret them—are an excellent illustration that even the simplest possible explosive source can create exceedingly complex seismic waves. I have used Leet's observations here to highlight a major cause of difficulties in discriminating between explosions and earthquakes by the seismic waves they create, that would plague nuclear test ban negotiations in the decades to come; that is, that the waveform features of a seismogram are far more due to properties of the earth along the path followed by the seismic waves than to the nature of the source. Although Leet was not thinking in such terms, difficulties in discriminating between earthquakes and explosions by their seismic waves could have been foreseen, even in the recordings of the first atomic test.

Regional Observations of *Trinity*

Seismic observations of *Trinity* at greater ranges than Leet's had been attempted by the Manhattan Engineer District, the agency responsible for developing the atomic bomb under the command of General Leslie Groves (Ziegler and Jacobson, 1995, pg. 38). Their purpose was to measure ground motions to be used to counter potential claims of damage to man-made structures by the explosion. Stations at El Paso, 210 km, Tucson, 430 km and Denver, 690 km from *Trinity* gave negative results (Figure 1.1). This report gave no information on the kind of seismometers installed, but we can infer from the results to be cited next that they were not sensitive, short period electrodynamic instruments of the type in use by institutions studying distant earthquakes.

Beno Gutenberg, Director of the California Institute of Technology's Seismological Laboratory—and the one who first gave me a glimpse into the science of seismology while a student at Caltech a year before—published on *Trinity* in October 1946. Gutenberg was a true giant in the field of seismology. He was trained in Germany in meteorology, and for his thesis, he studied microseisms—the ever-present, minute, vibrations of the earth's surface are found to exist everywhere if sufficiently sensitive seismographs are available to detect them. His scientific publications began in 1911 and shortly thereafter he made the first accurate determination of the radius of the earth's liquid core. He was a prodigious worker, with several books and several hundred scientific articles to his credit, some of which were published after his untimely death in 1960. As will be seen, his numerous other important areas of research led to discoveries and conclusions that pervaded our thinking on seismic wave propagation, seismicity (frequency and spatial distribution of earthquakes) and data interpretation in the early years of nuclear test detection.

On *Trinity*, Gutenberg reported that 10 seismic stations at distances between 437−1130 km from *Trinity* had recorded signals on short period vertical component Benioff seismographs. He remarked that, "Unfortunately, the time of the explosion is known only within about ± 15 seconds," a veiled critical reference to Leet's failure to record an accurate time for the *Trinity* explosion. (Much of what is known about the crust and deeper interior of the earth is derived from analyses of the time of travel of seismic waves along various paths downward into the earth, where they are reflected or refracted upward again. The time of origin is thus a crucial quantity in such studies. Since we can never know the time or location of an earthquake with high accuracy, a planned explosion represents a rare opportunity to obtain accurate knowledge of these parameters—an opportunity that no seismologist should overlook.)

Three stations, at 437 km (Tucson), 965 km (Palomar), and 1010 km (Riverside), had recorded the first arriving P-waves clearly. From these recordings, Gutenberg calculated the time of the *Trinity* explosion to be 11:29:21 GMT, which remains to this day the best estimate of the time. The successful recording at Tucson contrasts with the failure by the Manhattan project engineers to record seismic data at almost the same location, presumably with other types of seismometers. Gutenberg noted that S-waves were detected at seven stations, and that the beginning "was not very clear as is true in earthquake records from corresponding distances" —again hinting at difficulties in discriminating between earthquakes and explosions through their seismic waves.

The remainder of Gutenberg's article was about acoustic waves detected at the same stations. Gutenberg had done early research on low-frequency sound propagation through the upper atmosphere while he was still in Germany during World War I. Large explosions incident to the war were his sound sources. Sound waves, analogous to seismic P-waves, propagated upward into the stratosphere, and were refracted downward again at distances of hundreds of kilometers. In a course he taught at Caltech on physics of the upper atmosphere, he described long-range sound propagation—a major source of what was known about the structure of the upper atmosphere at the time—as "Seismology turned upside down."

Several years ago Dr. Kenneth Olsen of GCS International, Lynwood, Washington sent me a copy of a letter from the late Professor Otto Nuttli reporting that weak surface waves were recorded at St. Louis and Florissant, Missouri. More recently, Professor Robert Herrmann of St. Louis University has kindly sent copies of the seismograms to me. I can confirm that the signals were indeed weak; it seems possible that the waves were noticed at the time only because the analyst—Ross Heinrich according to Nuttli—had classified knowledge of the explosion. However, I value these recordings as rare mementos of *Trinity*, and the beginnings of nuclear test detection seismology.

Operation Crossroads

Operation Crossroads, designed to determine the effects of atomic explosions on naval vessels, took place in 1946 in the equatorial central Pacific at Bikini Atoll. For this purpose a fleet of obsolete U.S. and Japanese vessels ranging in size up to battleships were assembled in the Bikini lagoon. They were to be subjected to an atomic explosion in the air, a second under water in the shallow lagoon, and a third in deep water (later canceled because of extensive radioactive contamination caused by the second shot). In addition to the primary weapons effects studies, a Remote Measurements Project was included to help assess methods that might be used to detect possible Soviet tests.

The first explosion took place on 30 June 1946. The Crossroads *Able* test, an airdrop of 21 kt, was not detected by seismic stations in the U.S. Nor was it well detected by the Joint Army/Navy Remote Measurements Program at seismic stations closer to Bikini (Ziegler and Jacobson, pg. 54, 1995). Again, the reference gives no information on the kind of seismometers deployed.

The *Baker* test, an underwater shot of 21 kt at a depth of about 30 meters (m) took place on 24 July 1946. Gutenberg and Richter (1946) reported seismic signals at eight U.S. stations on the West Coast 69°—79° from the explosion.[1]

As had been the case for the *Trinity* explosions, these signals were recorded by short-period vertical Benioff seismographs. Figure 1.2, adapted from Carder and Bailey (1956), shows one of these recordings. Although the signal may seem small, note that a signal several times smaller would still have been detectable; i.e., its motion would have been noticeably different than what preceded it. Gutenberg and Richter comment: "Only longitudinal waves (P) appear; no transverse or surface waves were found." They noted that arrival times were 3–4 seconds earlier than predicted by the Gutenberg and Richter (1939) travel time tables and attributed this to the "exceptional structure of the Pacific Basin."

Amplitudes of P were "about equal to those of a shock of magnitude 5.5 at the same distance." Although the authors did not comment on it, the Gutenberg-Richter (1945) relationship between magnitude and energy[2] implied that the seismic energy was 1.5×10^{21} ergs, or twice the total bomb energy actually released! Since only a small fraction of the total energy release of the bomb could have gone into seismic waves, the discrepancy shown by the *Baker* test was enormous. This large discrepancy was later a factor in Gutenberg and Richter's (1956) downward revision of their energy/magnitude formula by a factor of several hundreds.

Stations operated by the U.S. Coast and Geodetic Survey (USC&GS) at Kwajalein and Wake Islands recorded *Baker*, but similar stations on Oahu and Midway Islands did not (Carder and Bailey, 1958). The authors give little information on the seismometers.

Thus, we can assess the nuclear test detection picture in mid 1946 as follows:

- On the technical side it had been found that air bursts were difficult to detect seismically at long ranges. However, other technologies were potentially applicable for detection of air bursts, including detection of sound waves through the air and the collection of airborne radioactivity; but their capabilities could not have been well known because so few tests had taken place. And underwater shots of about 20 kt could be detected seismically to great distances by sensitive short-period seismometers, but the radioactivity necessary to prove atomic origin would not be present in the air at similar distances.

- On the administrative side there had been some attempts at serious study of the detection problem by the U.S. Navy, USC & GS and other organizations. The Soviet Union's recent rejection of a proposal to place all atomic energy activities under international control suggested that means to detect foreign atomic tests might be needed by the U.S. However, the U.S. had no coherent program of research nor any organization with the mission to advance the state of the art.

Science, Not Spies: Creating a Long Range Detection Program

The stimulus to improve this situation came from Lt. General Hoyt Vandenburg, Director, Central Intelligence Group (of the U.S. armed forces, and the predecessor of the Central Intelligence Agency), who urged the concerned agencies of the government to form a Long Range Detection (LRD) group to study the problem and to formulate a plan of approach to accomplish nuclear detection. Eventually this took place, and in June, 1947 the committee reported (paraphrased, AFTAC, 1997):

> The objective of this (LRD) program is to determine the time and place of all large explosions that occur anywhere on earth, and to establish beyond all doubt whether any of them are atomic in nature.
>
> It appears that the instruments and methods needed to accomplish these several aims are available, actually or potentially, and possess adequate sensitivity.
>
> It is the belief of this committee that the objective of the long range detection program can be obtained as follows:
> (a) Locate large explosions by a combination of sonic, subsonic and seismographic methods;
> (b) Obtain samples of the explosion products by an aerial sampling technique as near the scene of the explosion as practical; and
> (c) By chemical and radiologic analyses of the products of the explosion, determine its nature.

As a result, on 16 September 1947 General Dwight D. Eisenhower, Chief of Staff of the U.S. Army, issued this highly classified order:

> MEMORANDUM FOR THE COMMANDING GENERAL,
> ARMY AIR FORCES
> SUBJECT: Long Range Detection of Atomic Explosions
> The Commanding General, Army Air Forces, is hereby charged with the overall responsibility for detecting atomic explosions anywhere in the world. This responsibility is to include the collection and analysis of the required scientific data and appropriate dissemination of the resulting information.
>
> In carrying out this responsibility, the Commanding General, Army Air Forces, will utilize to the maximum existing personnel and facilities, both within and without the War Department, will establish appropriate arrangements with other interested agencies for necessary assistance and will effect and maintain liaison with all participating organizations.
> /s/ Dwight D. Eisenhower

Note that this mission was to be carried out worldwide, and by scientific means (not by spying). It also encouraged the use of existing capabilities of other agencies of the Government, and gave the authority to make arrangements directly with such other organizations.

Eisenhower's order was issued just two days before the National Security Act of 1947 went into effect, reorganizing the structure of the U.S. military services. Responsibility for long range detection was transferred, along with most other Army Air Corps missions and facilities, to the U.S. Air Force, created as a separate service on 18 September 1947. In turn, the Air Force assigned this mission to the Air Material Command, the organization responsible for Air Force research and development. Within the Material Command, a highly classified organization known as AFMSW-1 was formed as part of the Special Weapons group, with offices in the Pentagon. Under Dr. Ellis Johnson, the organization's first technical director, AFMSW-1 began research on methods of bomb detection by means of seismic, sonic (acoustic), and radioactive debris collection methods. The program became known as the Long Range Detection (LRD) project.

Conflict Between Committees of the Research and Development Board

The same National Security Act that made the Air Force a separate service, also established a joint Research and Development Board (RDB). Dr. Vannevar Bush, President of the Carnegie Institution of Washington, D.C., was appointed its chairman, with two representatives each from the Army, Air Force, and Navy. Bush was perhaps the inevitable chairman, having served during World War II as President Franklin D. Roosevelt's scientific advisor and as director of the Office of Scientific Research and Development. In that office, he had presided over the development of radar, the proximity fuse, penicillin, and numerous other scientific developments—many accomplished in close cooperation with the British—that helped shorten the war. The newly created RDB had cognizance over all military Research and Development.

Twelve committees were established, including two of particular relevance to the LRD project:

First, a Committee on Atomic Energy initially chaired by Dr. James B. Conant, of Harvard University, and including Dr. J. Robert Oppenheimer, the "father of the A-bomb"; General Leslie Groves, the commander of the Manhattan Project that built the A-bomb; and six officers of the Military Liaison Committee (the formal body that transmitted military requirements to the Atomic Energy Commission to guide their weapons developments).

Second, a Committee on Geophysics and Geographics with Dr. Helmut Landsberg of Air Force Cambridge Research Center as Director, and other members chiefly from the three military services. This committee had a subordinate Panel on Seismology, with members chiefly from academia, that assisted in the review of seismic aspects of the LRD research.

Both committees of the RDB claimed oversight responsibilities over the LRD program, or at least major portions of it. The efforts of the two committees to exercise oversight inevitably led to rivalry and conflicting guidance to AFMSW-1, as well as conflicting advice to the higher military authorities. The Atomic Energy Committee seemed to have an advantage in exercising this responsibility, perhaps because of the eminence of their leadership in military developments during World War II. Whatever the reason, this committee became skeptical about the probability of success, and even the need for, seismic and sonic methods, and their advice substantially impeded the LRD program.

On 22 December 1947 the Committee on Atomic Energy dealt a blow to the LRD program, noting "that a view has been expressed that techniques and instruments, etc. are available potentially or actually for the long range detection of nuclear explosions," however, "This Committee feels that there is grave doubt whether this optimistic view is justified."

The Atomic Energy Subcommittee went on to warn that radioactivity monitoring "may be of limited utility." These statements reflected the views of a number of noted physicists, apparently including James B. Conant and J. Robert Oppenheimer, that LRD might well be impossible.

The view at the time among the skeptics was that only surface or near-surface bursts such as *Trinity* would produce radioactive fallout. Their belief was that at higher altitudes all components of the bomb would be completely atomized and dispersed. Thus, there would be no radioactive particles large enough to be collectible by filtering the air, preventing the acquisition of the radioactive samples needed to prove that an atomic explosion had taken place. Even near-surface bursts that ingest large quantities of dirt particles which then become the carriers of radioactivity, might be made much less detectable by detonating in the rain, which would wash out most particulate matter near the test site. And although underground and underwater tests might be detectable, the absence of radioactivity would prevent establishing nuclear proof. The focus of the U.S. was very much on detecting the first Soviet test, and it was considered possible, if not likely, that the USSR would test under conditions designed to reduce detectability.

In spite of the doubts expressed by the RDB, AFMSW-1 pressed on with its mission. One action in April 1948 was to form a seismic advisory panel chaired by Dr. Roland F. Beers, president and founder of The Geotechnical Corporation, a successful petroleum exploration company in Dallas, Texas. The membership list of the panel read like a "Who's Who of American Seismologists;" virtually every American institution concerned with classical seismology was represented.[3]

That panel included Professor Perry Byerly, under whom I did my graduate work at the University of California. I remember this from appearances before the panel in the early 1950's. But I also recall his "proof," associated with his work on the panel, that it was always dark in Kansas: after all, he had traveled between Berkeley and Washington, D.C., countless times to attend panel meetings, and he had never seen daylight in Kansas! Perhaps a typical Byerly story to emphasize a lesson in science: in this case that statistical methods are no substitute for physical reasoning. He was a strong believer in observational seismology—"the data's the important thing," he would say. "Don't worry about whether it fits the theory—the theory will follow the data."

Byerly was a pioneer in developing the science of seismology in the United States. Starting with the first two seismographic stations in the Western Hemisphere (at Berkeley and Mt. Hamilton, California), he built a network of stations capable of keeping track of earthquakes throughout central and northern California. But he was quite at home with theory as well. His early research on deducing the motions at an earthquake focus from observations of the direction of the first motion of the P-waves would later become the physical foundation for one of the first criteria for differentiation between seismic signals from earthquakes and explosions.

A Breakthrough for Long Range Detection Methods

Operation Sandstone, a weapons-development test series to take place early in 1948 in the Pacific, was defined as a critical test of Long Range Detection methods. It included atomic tests of 37, 49, and 18 kt, all on 60 m towers at Enewetok during April–May, 1948. William Cloud, of the U.S. Coast and Geodetic Survey, and Aaron Heller, of the Naval Ordnance Laboratory, placed seismographs on 11 islands in the Pacific. They reported no detection beyond 500 miles—a discouraging result for seismology.

However, Operation Sandstone did result in a key discovery concerning airborne radioactivity. Analysis showed that the radioactive debris collected at great distances contained not only contaminated soil particles, but also microscopic metallic spheres of radioactive materials. These spheres could only have been formed by condensation from the bomb plasma, just as water droplets coalesce from water vapor to form clouds and rain. This finding proved that the material had not been simply atomized and dispersed uncollectably within the atmosphere, as skeptical scientists on the Committee on Atomic Energy had thought likely. This discovery was an important boost to the standing of AFMSW-1 in scientific circles, and enhanced its ability to continue with its research program.

With the good news about the detectability of radioactivity, Dr. Ellis Johnson, the Technical Director of AFMSW-1 "declared success" for the research, and he moved rapidly to form an operational "Interim Monitoring System" for collecting airborne radioactivity. The main component of the system was to be operated by the Air Weather Service of the U.S. Air Force. This called for equipping aircraft flown by the Air Weather Service with devices to filter particles out of the air while flying routine meteorological missions between Japan and Alaska, and from Alaska to the North Pole and back. Each filter would be turned over to laboratories operated under contract to AFMSW-1 for analysis after each flight. Arrangements were soon made to have similar flights by the British Air Force covering the North Atlantic region. The U.S. Navy also collected rainwater at a number of ground stations, and analyzed the water for possible radioactivity washed out of the clouds.

Under Johnson's concept, seismic and sonic (acoustic) systems were to be added to the Interim System after such systems were further developed. Meanwhile he called for the operation of experimental seismic and sonic networks, chiefly in the U.S., to test new concepts, and to train people for the final system. This pattern of establishing experimental systems that would ultimately "graduate" into operational systems became the model by which the Interim Monitoring System was to grow for several years to follow.

During the midsummer of that year, to reflect its emerging operational role, AFMSW-1 was redesignated as the Air Force Office of Atomic Energy, "AFOAT-1," a name it would bear for the next ten years. In August, 1948, AFOAT-1 submitted a budget for research and development during Fiscal Years 1949 and 1950, but the Committee on Atomic Energy would approve only that part devoted to work on nuclear debris collection and studies to determine if an explosion could be distinguished from an earthquake. AFOAT-1 appealed, asking for a review by the Committee on Geophysics and Geographics. The Board, however, supported the Committee on Atomic Energy.

According to Northrup and Rock (1962), "From then on AFOAT-1 was engaged in a continuous running battle with the Committee under Oppenheimer and then under James Conant." About the same time, Johnson resigned as technical director, frustrated by his inability to obtain RDB approval for work that he considered vital to AFOAT-1's mission.

After a short interim period, in September 1948 Doyle L. Northrup, the Deputy Technical Director, replaced Ellis Johnson as Technical Director and he became the guiding genius of the developing new system for several decades thereafter. This comment is not to minimize the role of the exceptionally high caliber military commanders of AFOAT-1 or its successor agency, AFTAC, during Northrup's tenure as technical director. Their command and management accomplishments were great, and essential to the success of the agency. But the underlying theme here is technical, and in that domain Doyle Northrup was the central force. His prior career had been in high-energy physics research at the Massachusetts Institute of Technology, followed by research and development work on torpedoes for the Navy. Working in the latter capacity, he had been present in Pearl Harbor during the Japanese attack on 7 December 1941. In his work on torpedoes he made major contributions to finding and correcting a defect that prevented detonation when torpedoes impacted their target, and for this he received the Navy's Meritorious Service Award.

The Boner Panel

Committees of the RDB continued to exercise oversight of AFOAT-1's program without a clear resolution of the sometimes-conflicting advice by the two committees. Then, in an apparent move to improve the situation, responsibilities were vested in the "Boner Panel," formed in late 1948, to evaluate AFOAT-1's plan and funding request for the LRD Program. The panel was to report to RDB through both the Atomic Energy and the Geophysics and Geography committees. Charles P. Boner, physicist of the University of Texas chaired the panel, which also included seismologist Father James B. MacElwane, S.J., of St. Louis University, James B. Fisk, at that time Professor of Applied Physics at Harvard University, Joseph Boyce, physicist from N.Y. University, and meteorologist Athelstan Spilhaus, of the University of Minnesota.

On 12 September 1949, the Boner Panel reported the results of its detailed review of the AFOAT-1 plan. Their report was critical of the Air Force's emphasis on detecting a first Soviet test, recommending greater emphasis on monitoring the overall atomic capabilities of the USSR, including uranium mining, fissionable material production and so on. Other points were:

- The first Russian test was likely to be on a tower
- An underground test would be the "worst case" for detection
- Budget cuts should be made for the seismic and sonic programs
- The Joint Chiefs of Staff (JCS) should "reevaluate the necessity for de
tecting a foreign atomic explosion by instrumental means, with a view to canceling the R & D programs of LRD."

One can wonder at the wisdom of cutting the research budget on the only method of detecting "worst case" tests, and marvel at the group's temerity for implicitly suggesting that the JCS embrace some form of espionage in the USSR. However, unknown to the Boner Panel, the Soviet Union had already conducted its first atomic test.

Detecting *Joe-1*

By September 1949, initial developments in radioactivity, seismic and sonic methods had been incorporated into AFOAT-1's Interim System. However, at this time there was no dedicated network of operational seismic stations. Instead, data from a selected set of seismic stations at academic institutions and governmental agencies around the world were discretely collected via telegraphic reports and made available at AFOAT-1 Headquarters. There, Dean Carder, Chief Seismologist of the USC&GS and his colleague, Leslie Bailey, reviewed the data on behalf of AFOAT-1, searching for potentially suspicious seismic events in the USSR. Dean Carder had been an early student of Perry Byerly. After receiving his Ph.D., he studied the vibrations of man-made structures during earthquakes. He also wrote classic papers on the influence of reservoir water levels in triggering local earthquakes, using data from stations near Boulder Dam (Hoover Dam) as the reservoir slowly filled after construction.

On 29 August, 1949 the first Soviet test, which they called *"First Lightning,"* as we discovered more than four decades later, and which we dubbed *"Joe-1"* (after Joseph Stalin), was detonated on a 50 m tower south of Semipalatinsk, Kazakhstan. No seismic signals were detected, nor were acoustic signals noted at the time. However, an Air Weather Service "Loon Charlie" flight from Misawa, Japan to Eielson Air Force Base, Alaska, on 3 September 1949 collected samples measuring roughly 100 "counts per minute" (on a Geiger counter), more than twice the normal background level. The filter papers were rushed to AFOAT-1's contractor, Tracerlab, for analysis, and AFOAT-1's Data Analysis Center in Washington issued "Alert 112." The Alert, in turn, triggered more flights, and on 5 September a flight from Guam to Japan collected a sample measuring 1,000 counts per minute—more than twenty times the normal background level! Samples were also sent to the Los Alamos Scientific Laboratory for analysis, and it was clear to the scientists involved that there were large amounts of fresh nuclear debris in the atmosphere. The cause of this was less clear, since most intelligence estimates placed the earliest possible Soviet bomb test in 1951 or later. The possibility that the debris had come from a reactor accident had to be considered.

By 9 September radioactive fallout was also recorded by six ground filter stations operated by AFOAT-1 and by the U.S. Navy at Kodiak Island in the Aleutians. By 10 September laboratory analysis had shown that the debris had come from the fissioning of plutonium, and that this had occurred between 26–29 August. A few days later, radioactive debris reached the North Atlantic, where it was collected by cooperating U.K. aircraft. Samples of the debris were sent to laboratories within the U.K. for analysis.

By then, laboratory analyses had confirmed that the source had been an explosion of a plutonium device, most probably the first Soviet test. Ironically, amidst frantic collection and analysis activity, "A clowning fate picked this juncture" (Northrup and Rock, 1962) for the Boner Panel report, calling for cancellation of research on LRD to be presented to the Atomic Energy Committee. It was forwarded rapidly to the RDB, and as a result, a "Stop Work" order was issued to AFOAT-1 by the Board (at that time chaired by Karl T. Compton, President of MIT). As recounted to me by Doyle Northrup, the order was received while he and his colleagues were busily plotting curves of "counts per minute" as a function of time, showing decay rates typical of radioactive bomb products. The order was promptly canceled when the RDB learned that what AFOAT-1 had detected was probably a first Soviet test.

President Truman was kept informed of the news as accumulating data made AFOAT-1's conclusion increasingly more probable. Truman was reported to be skeptical that it was actually bomb debris, and he made no public announcement, in part pending an independent validation of the conclusion. That came on 19 September from a panel of high level physicists, chaired by Dr. Vannevar Bush and including such eminent scientists as Dr. J. Robert Oppenheimer, Director of the Los Alamos Scientific Laboratory at the time the first A-bombs were built, and Dr. Robert Bacher, also a developer of the first bombs, and later an AEC Commissioner. They confirmed, in unambiguous language, that a Soviet nuclear test had occurred, and that it was a close copy of the U. S. plutonium bomb.

Joe-1 was a great surprise to the intelligence community and to most U.S. scientists in atomic energy circles! Most believed that the USSR would not be capable of such a test for at least a few more years. Several scholars have pointed out the probable role of espionage in this early success. See, for example, Michael S. Goodman's recent reassessment of the significance of the information on the US/UK/Canada atomic program passed to the KGB by Klaus Fuchs (Goodman, 2007). Fuchs was arrested in England only a few months after *Joe-1*. He freely confessed to providing detailed atomic weapon design information to the Soviets while working in the theoretical division at Los Alamos before and after *Trinity*, and he continued passing information after returning to England in 1946. Fuchs was convicted and imprisoned in 1959.

A surprise was also in store for the Soviet Union: On 23 September President Truman and the Prime Ministers of Canada and Great Britain simultaneously announced that an atomic explosion had taken place in the Soviet Union. (It is now clear that the Soviets had believed that the test had taken place in complete secrecy and was undetectable outside the territory of the Soviet Union; see Appendix B.)

I should remark here that the several previous paragraphs discussing the influence of the RDB on the AFOAT-1 research and development efforts is intended only as a hint of what actually went on. For much more detail the reader should refer to Northrup and Rock (1962) and to Ziegler and Jacobson (1995). Suffice it to say that AFOAT-1 staff members had succeeded in accomplishing their initial goal only through their own initiatives and dedication to their mission, and not as a result of beneficial guidance by the eminent scientists on committees of the RDB.

TWO

Genesis of a Seismic Detection System

Research at Beers and Heroy

Success in detecting Joe-1 demonstrated that those skeptical about detecting atomic tests at long ranges were wrong, and it removed much of the resistance by committees of the RDB to further research. This extended to the sonic and seismic techniques, even though they had played no part in the initial success. For development of the seismic techniques, work was authorized by the USC&GS, Naval Ordnance Laboratory, Office of Naval Research, The Geotechnical Corporation of Dallas, Texas, and the consulting firm of Beers and Heroy (B&H) in Troy, N.Y. Following are some of the steps that ultimately led to an effective seismic system.

My own direct involvement with nuclear test detection began in mid-September, 1949 , one week before the President's announcement of Joe-1. At the time, I had completed most of the requirements for a Ph.D. at the University of California in Berkeley. I was married and had a young daughter. However, I had used up my eligibility for educational support through the "G.I. Bill of Rights;" my savings from three years service in the Navy were almost depleted; and my earnings as a Research Assistant, by themselves, were insufficient for our family's needs. Dr. Byerly had recently attended a review of the B&H program, and discovered that there was no seismologist on the B&H staff, even though seismology was the primary focus of their research. The solution to both problems was made obvious to me by a visit from Dr. Roland F. Beers -- I should defer my university studies and assist Beers & Heroy while saving funds for my final work at Berkeley. Dr. Beers could not tell me much about the work, except that it was secret, but he assured me, as did Byerly, that it would be challenging and important to national security. Shortly afterward, my family and I departed for upstate New York.

As I learned while waiting for a security clearance, Roland Beers had founded The Geotechnical Corporation in 1936, and developed it into a successful geophysical prospecting organization. At that time, American demand for petroleum products was rapidly growing, but known oil fields and potential new fields having visible surface geological expressions had all been explored. Methods for "seeing" deep rock structures that might hold petroleum were being developed at the time, and Beers was among the early entrepreneurs in employing geophysical methods to locate promising locations for deep drilling. These methods included gravity, magnetic and electrical measurements of the earth, and most importantly, measurements of seismic waves created by small dynamite charges reflected and refracted from deep rock layers.

In addition to his commercial activities, Roland Beers had served on several prestigious panels and advisory boards for U.S. Governmental agencies, and as previously noted, was Chairman of the Air Force's seismic advisory panel. He was also a Professor at Rensselaer Polytechnic Institute. The Geotechnical Corporation ("Geotech") operated a number of petroleum exploration crews and their associated prospecting gear under contract to various oil companies. Support to these operations was provided by laboratory and maintenance facilities at the headquarters on Haggar Drive in Dallas, Texas.

Dr. William B. Heroy, Sr. had come to Geotech after a long career as a geologist associated with oil exploration, as well as in advisory positions with the U.S. Government on petroleum resources during World Wars I and II. "Senior," as we commonly referred to him, had been Chief Geologist with Sinclair Oil Company before World War II, and he met Roland Beers while Beers was doing geophysical work for Sinclair. During the recent war Senior worked in Washington in the Petroleum Administration for War. After the war, Beers offered him a position as Vice President of the Geotechnical Corporation, and effectively turned the operation of the company over to him.

Shortly thereafter, the Beers & Heroy partnership had been formed to conduct business with the government, which involved procedures and legal considerations that differed from those normal to Geotech's commercial business. The firm of B&H was administratively supported by The Geotechnical Corporation in Dallas, but operated from an office in downtown Troy, N.Y. and laboratory facilities on Beers' estate at nearby Pinewoods Avenue.

Day-to-day activities at Troy and Pinewoods were directed by James M. Klaasse, a physicist who had worked on the development of airborne magnetometers at the Naval Ordnance Laboratory to detect submarines during the war. After the war, while at the W. & L.E. Gurley company in Troy, N.Y., he worked on airborne magnetometers with Beers, for use in petroleum exploration. Beers had arranged with the Gurley Company for Klaasse to take a leave of absence to manage the operation of B&H. He was thus my supervisor, and he became my mentor on engineering aspects of seismic instrumentation. Although I had studied the theory of seismographs, and helped repair and maintain operational seismographs under Professor Byerly, I soon discovered under the tutelage of Jim Klaasse that there was much more to learn from an engineering point of view.

The research and development project being carried out by B&H was funded through AFOAT-1, and was conducted under the direction of AFOAT-1 staff members Ben S. Melton and his supervisor, J. Allen Crocker. Ben Melton was an electrical engineer, whose early career was involved with instrumentation for petroleum exploration. Subsequently, in 1930, he became party chief of the Geophysical Service Corporation's seismic exploration crew 311. He was one of the first scientific workers for AFOAT-1 starting in 1948. J. Allen Crocker was also one of the earliest scientific staff members of AFOAT-1. He had been present at the *Trinity* explosion, where he operated seismic equipment in association with L. Don Leet. He was an Assistant Technical Director of AFOAT-1, with responsibilities to develop seismic and electromagnetic-pulse methods for detecting bombs.

As I learned after receiving my security clearance, three principal activities were going on at B&H in the fall of 1949:

- Operation of several seismic stations; seven were planned to the best of my recollection, to be distributed in a north-south direction along the Hudson Valley and in an east-west direction along a line from Troy to the Harvard College Observatory seismic station. A station at Rensselaer Polytechnic Institute (RPI) in Troy was to be common to both lines. This station network was never completed, although stations were operated at RPI, at Pinewoods Avenue, at the Harvard Observatory, and near Grafton, N.Y.
- A study of the generation, propagation and detection of seismic waves, primarily from the standpoint of a literature study. This work, as well as analysis and evaluation of the seismograms from the four stations, was conducted in the Troy office under my direction.
- Engineering test and evaluation of the equipment being operated at the seismic stations, which was initially the principal work of the laboratory at Pinewoods Avenue.

The Unsuitable Seismographs

I was puzzled when I learned that the sensors at the seismic stations consisted of two horizontal seismographs sensitive to waves with periods of several seconds or more, and a vertical seismograph sensitive to waves with periods of one to several tenths of a second. Three component instruments permitted measurements to be made in three mutually perpendicular directions, to fully define the motions of the Earth's surface. However, the measurements from these instruments could not be directly combined since the vertical would be sensing motion at different periods than the horizontals. As a result, instead of measuring amplitudes of coherent motion in different directions—key information in determining types of waves—one would be measuring relative amplitudes at different frequencies.

Common seismological practice was (and is) to orient the instruments to measure earth motions in the East-West, North-South, and up-down directions. The horizontals had been designed by the David Taylor Model Basin of the U.S. Navy and were referred to as the "DTMB," or "Reed" (after the manufacturer) instruments. These instruments employed capacitance-bridge transducers to convert ground motions into electrical signals and electrical feedback for stabilization and adjustment of the natural period of oscillation (about ten seconds). The vertical seismograph with a natural period of oscillation about one second was referred to as the "Stanley" (after the manufacturer). It had been designed for the Department of Terrestrial Magnetism of the Carnegie Institute of Washington, D.C. The Stanley employed a moving-coil-and-magnet transducer to convert ground motion into electrical signals, making use of a powerful magnetron magnet, surplus from the recently ended war. It also used an unconventional, non-linear, spring suspension that was intended to achieve the desired natural period without requiring the large vertical dimension of a conventional coiled spring suspension. Both types of seismograph used vacuum tube amplifiers and Brush pen-and-ink recorders. Further details are contained in Ben Melton's (1981) article.

Neither of these instruments had much in common with seismographs successfully operated by seismologists for many years. Typical observatory seismographs employed a mass suspended on a spring for measuring vertical motion, and on a hinged horizontal pendulum (we referred to these as "garden gate" pendulums) for measuring horizontal motion. A coil of wire was attached to the mass, which was suspended in the field of a magnet attached to the frame. When the earth moved, the inertia of the mass caused it to lag behind the motion of the earth. Differential motion between the coil and the magnet generated a weak electrical current (see Figure 2.1). This current caused movement of a ballistic galvanometer bearing a mirror, which deflected a narrow light beam onto photographic paper mounted on a cylindrical drum. As the drum rotated, sideward deflections of the beam traced the movement of the earth as a function of time. Such instruments were simple, stable, dependable and sensitive enough to record motions as small as the ambient noise level of the earth—excellent instruments for essentially unattended observatory operation.

The task of evaluating the operational performance of equipment for potential use by the U.S. Atomic Energy Detection System, or "AEDS," as it came to be known, fell chiefly to the staff of the Pinewoods Laboratory. Much of this work was under the immediate direction of Frank Snell. A difficult early task was to design and construct a "shaking-table" on which seismometers could be placed and subjected to precisely known small motions. The objective was to measure the seismometer's output in response to known inputs similar to motions of the earth caused by seismic waves. The device was used to calibrate seismometers at different frequencies, and especially to determine the linearity of their response to motions as small as a few millionths of a millimeter. Other work at the laboratory included trouble-shooting instruments from the field installations, in the attempt to improve operational reliability. Jack Hamilton, a recent engineering graduate of George Washington University, was a key participant in this design and construction of the shaking-table, and ran the tests on instruments under study.

During the period 1949–1950, this work in the laboratory together with the field operations demonstrated that the DTMB and Stanley instruments were unsuitable for surveillance of the USSR. They were costly, requiring full time operators in attendance at all times, and they required expensive batteries to obtain sufficiently noise-free low-frequency power supplies. Worse, they were unstable; the Stanley was never sufficiently stabilized to permit extended observatory type operations, and the DTMB required frequent adjustments by an operator to re-center the drifting pendulums. This latter problem was well illustrated when the great earthquake of 15 August 1950 occurred in Assam, India, at the eastern end of the Himalaya Mountains. As the seismic waves from the magnitude 8.6 quake rolled across that part of New England, station operators, to a man, turned off their seismographs, logging extreme instability as the reason! Large side-to-side motions of the recorder's pens were a common occurrence as far as the operators were concerned and the quake was mistaken for just another example.

A more fundamental problem for the DTMB was that it did not respond well to periods near one-second where I knew the teleseismic P-wave signals from atomic explosions were best detected, as Gutenberg's results had demonstrated from *Trinity* and Bikini *Baker*. The DTMB was much more suited to recording the 5–10 second microseisms from large storms at sea, which must have been why the Navy built these instruments in the first place.[4]

How or why the DTMB and Stanley instruments became the initial candidates for operation by the AEDS remains a mystery to this day. Melton (1981) suggested that seismological observatory instruments in use by universities and the USC&GS were not suited to rapid manufacturing. Billy G. Brooks, an early member of the B&H staff, has suggested recently (personal communication) that the instruments were readily available as surplus equipment in some Navy warehouse, and thus free to AFOAT-1. Perhaps so. But I also recall a statement by the Director of the Naval Ordnance Laboratory, an important participating agency in the seismic research at that time, that standard observatory seismographs were unsuitable because their response was not flat with respect to the displacement of the earth's surface, or to some other physically simple quantity like velocity or acceleration. (The statement reveals a lack of knowledge of the characteristics of the ambient noise in the earth, as will be discussed more fully in Chapter 5.) There may also have been a preference for "modern" pen-and-ink recorders, rather than the customary photographic recorders. But such recorders required the use of vacuum-tube amplifiers that, as was being discovered, were noisy at the low frequencies and extremely low power levels generated by seismic waves at quiet sites. Regardless of the reason for initial selection, these instruments proved to be unsuitable and the Pinewoods laboratory staff soon turned toward other instruments.

Selection of an Effective Seismograph

Because of its demonstrated success in bomb detection in 1945 and 1946, the Benioff short period seismograph became a prime candidate. Although the Stanley instrument had been unsuitable for observatory operation, we had been able to use it over intervals long enough for measuring the spectrum of short period noise in the earth. Using one-octave bandwidth filters (Figure 2.2), we found that the amplitude of the ambient seismic noise during quiet times at the Pinewoods and Harvard seismic observatories varied approximately as the third or fourth power of the period between about 0.5–5.0 seconds (Romney, 1953).

When this result was shown to Jim Klaasse, he immediately commented that this was crucial data for the selection of our seismograph, and that it strongly favored an instrument like the Benioff. Because the Benioff's response varies *inversely* as the third power of the period in this range, the product of the ground motion amplitude and the seismograph magnification is nearly constant, giving a flat noise output over a broad bandwidth ("white noise") optimal for detection. Benioff instruments were borrowed from Columbia University and tested extensively in the laboratory, including shaking-table tests; and they were selected, first for field-testing during the approaching Operation Greenhouse and ultimately for operational use. At Klaasse's request, Richard Arnett and others of the Geotechnical Corporation made contractual arrangements with Professor Benioff to manufacture a number of seismometers for the use of B&H.

Hugo Benioff, Professor of Seismology at Caltech, had invented the "Variable Reluctance Electromagnetic Pendulum Seismograph" in the early 1930's. Benioff's early scientific work was in astronomy, and he had assisted A. A. Michelson in his technically demanding work in measuring the velocity of light. This had been accomplished by sending a light beam 22 miles from Mt. Wilson to Mt. San Antonio and back, (in the San Gabriel mountains just north and east of Caltech), where it was sensed and timed by an elegant detector. The result was a significant improvement in the accuracy of this fundamental constant.

Benioff continued his interest in highly sensitive instrumentation as he moved into the field of seismology, and numerous innovations in seismometry are attributable to him. His variable reluctance seismometer, which we referred to as simply "the Benioff," had been designed to solve a long-standing problem: how to achieve high sensitivity at short periods of the order of a fraction of a second to about one second. It was known that the microseismic noise was low in this period range, and that P-waves from distant earthquakes could be observed at these periods. Previous attempts had foundered on the inability of seismometers to generate sufficient electrical power to damp the instrument while sensing these weak signals; the variable reluctance transducer solved this problem. As we have seen from a few examples, this is also the range of periods where the bomb signals may be detected at great distances.

Benioff had broad interests in seismology, and his work on deep-focus earthquakes led to the definition of what we now call "Benioff zones," in regions of the earth where one crustal plate dives under another. He was also an accomplished musician, and had debuted as a concert pianist. He became a consultant to the Baldwin Piano Company. When I visited his home a few years after the selection of his seismometer for a realistic operational test, he explained that he had miniaturized the variable reluctance transducer, in order to attach it directly to the strings of pianos and other stringed instruments. The objective was to obtain purer tones from the instruments to be used for making improved musical recordings.

Seismometer Arrays

Our understanding of the nature and characteristics of seismic noise — the ever-present, minute vibrations of the earth that limit detection of seismic events — was rudimentary at the time. It was known that relatively strong waves in the 5-to-15-second band were generated by storms at sea, and that they propagated at low velocities along or near the surface; but just how water waves at sea produced surface waves on the solid earth was not clearly understood. It was also known that noise at shorter periods, near one second, could be created by surf, wind, and activities of man; but their propagational characteristics had not been studied to any significant degree. The stronger noise from storms could be reduced or eliminated by design of the seismograph to filter them out, but the residual noise at short periods would still limit the detectability of bomb signals near one-second period.

However, exploration geophysicists had found that, if the wanted signal wavelength is effectively much longer than noise wavelengths at the same period, seismometer arrays can be used to enhance the signal relative to the noise. Profiting from earlier research by oil prospectors, we developed a simple theory for noise-reducing seismometer arrays, based on the unproven assumption that signals and noise near one Hz behaved as they did at much higher frequencies for prospectors. Prospectors used arrays (a group of seismometers; see Glossary) to suppress "ground roll," their principal form of noise. In placing their seismometers, prospectors exploited the fact that the direction of travel of this noise was known because the explosion they had fired generated most of it. They placed their sensors along a line toward the shot, spaced so that a given wave traveling along the ground would arrive at each sensor at a different time. Thus upward motion might be detected by one sensor at the same time as downward motion was present at another sensor. Summing the outputs would result in partial cancellation of the noise. The prospector's signal of interest arrives at a steep angle from below, so it arrives and is sensed almost simultaneously by each instrument, and thus is added constructively. The result is to increase the ratio of signal to noise.

As we discovered, earthquake seismologists do not generally know where the short period noise is coming from, and, indeed, it may come from many directions simultaneously; so a different strategy of seismometer deployment was needed. Since digital, delay-and-sum technologies (used in modern arrays) were still a decade away, our operational concept was to place the instruments in a straight line oriented broadside to the direction of the expected signal sources, and directly sum the electrical outputs. The expected result would be that the wanted signal would arrive simultaneously at each sensor and add coherently, but the noise would arrive in random order at the sensors, resulting in partial noise cancellation. In the absence of amplifiers suitable for stable, low-noise application at the low frequencies of seismic waves, techniques for summing and recording the signals directly from the seismometers, while maintaining proper isolation and damping were developed by the Pinewoods Laboratory group.

Meetings with the Boner Panel

Results of the work at Troy and Pinewoods were periodically reviewed by AFOAT-1 staff members, and from time-to-time presented to the Boner Panel for final approval. On occasion, Jack Hamilton and I traveled to Washington D.C. to support Roland Beers in such presentations. We stayed as guests at the Cosmos Club—a prestigious club whose members were predominantly senior scientists and other intellectuals through Bill Heroy, Sr.'s influence as a member. To us, the place became known facetiously as "the Old Duffers' Club," since we were always at least one generation younger than any other guest. But we truly enjoyed staying there, savoring its nineteenth century opulence and genteel Southern-style service.

The Boner panel meetings held in various highly secure facilities around Washington must have been held jointly with the AFOAT-1 seismic panel (and perhaps other panels). They were a welcome opportunity for me to keep in contact with Perry Byerly during my leave of absence from graduate school. And most appreciated, it was also a forum where I could discuss our work openly with senior scientists like Beno Gutenberg and Father James B. MacElwane, S. J.—the founder of modern seismology at Berkeley while on a two-year leave of absence from his religious order. Such discussions, of course, were not possible at the normal unclassified scientific meetings. I have only one declassified report of these meetings, but it is clear that our work on instrumentation and arrays was discussed with the Panel members, and their report of 1 July 1951 explicitly endorses the array work.

Testing the Experimental Seismic System

The first operational seismometer arrays for the detection of teleseismic P-wave signals were constructed in early 1951 to participate in a realistic operational test, attempting to detect nuclear tests of the approaching Operation Greenhouse in the Pacific. Station sites were selected for installing four-element "earth-powered" arrays several miles north of College, Alaska (near Fairbanks), near Ankara, Turkey, and at three sites in Wyoming. The Wyoming stations formed a triangular, "tripartite network," with dimensions of the order of 120–200 km. One station was located at Pole Mountain, 30 km west of Wyoming's capital, Cheyenne, and about 20 km east of Laramie; a second near Encampment, 120 km further west and just north of the Wyoming/Colorado border; and the third several miles south of Douglas in East-Central Wyoming. (See Figure 1.1.)

The earth-powered arrays used short-period vertical Benioff sensors, mounted on concrete piers poured on bedrock. North-South and East-West oriented horizontal Benioffs were also installed on one pier. We constructed small wooden or concrete block shelters around the piers, to give some protection from wind and thermal changes. Electrical signals generated by each seismometer were transmitted to a central recording station over a "Spiral-4" cable[5] laid along the ground. Within the central recording stations, signals were recorded individually from each sensor, and the signals from the four vertical instruments of the array were also summed and recorded on a 35 mm film recorder.

Arrays at Encampment and Douglas, Wyoming were installed and operated by Beers and Heroy personnel. Arrays at Pole Mountain, Wyoming, and in Alaska and Turkey were operated by AFOAT-1 military teams. B&H also deployed civilian teams to the Southwestern Pacific islands of Mindanao, Truk, and Yap to operate 3-component stations for Operation Greenhouse during April–May, 1951.

W. B. Heroy, Senior liked to tell stories, and his story about these remote operations became a classic within the inner, classified, circles conducting the program. According to him, the young operator of the seismographs on Truk noticed an impending failure of a recorder drive motor. After preparing a message requesting a replacement motor, he delivered it to a member of the U.S. Navy communications detachment supporting scientific operations on Truk. The sailor who was to transmit the message asked what priority should be assigned. "I don't know," responded the operator, "what's available?" "Well," said the sailor, "Routine is normal, but it can be slow." "What's the fastest?" asked the operator. The sailor responded, "That would be Flash." "OK, send it Flash."

The military normally reserved a Flash priority for highly significant matters like initial hostile enemy contact. Sending one would clear all Pacific communications channels of any other traffic. Within minutes a copy was on the desk of a puzzled aide to President Truman. It was also on the desk of the commander of an Air Force base near Dallas.

Sirens howling and lights flashing, the Air Force base commander drove up to the Geotechnical Corporation's office. He delivered a copy to Senior, who signed a receipt for it and read it calmly. Thinking some momentous matter must be encoded in the message, requiring urgent response, the commander blurted out, "Well, what are you going to do about this?" "I'll send him the motor," Senior calmly replied.

An Operational Seismic System.

The *George* shot of Greenhouse, a 225 kt explosion on a 60-meter tower at Enewetok on 8 May 1951, was successfully detected at the deployed seismic stations out to a distance of more than 9000 km. The *Dog* explosion of 81 kt, also on a 60 meter tower was also detected, but less clearly. The signal amplitudes were minuscule: in the range of a few millimicrons, or a small fraction of a wavelength of light. Such small signals are entirely undetectable at most seismic stations; but the experimental arrays had been placed at carefully selected sites where short-period noise level was exceptionally low. With the success in recording Greenhouse shots, this time it was Doyle Northrup who "declared success" for the seismic system. The three seismic stations in Wyoming as well as the stations in Alaska and Turkey were each designated to become operational parts of the AEDS at this time.

The station at Pole Mountain, located on a military base (Francis E. Warren Air Force Base), had been constructed with the idea in mind that it might become an operational station. That was not the case for stations near Encampment and Douglas, which had been minimally constructed for the Greenhouse experiment. It was thus necessary for Beers and Heroy to undertake a rapid "rehabilitation" program, improving access roads, constructing new and permanent instrument piers and shelters, laying new cables, installing electrical power generating equipment, constructing a service building with temporary living quarters, and other work to make the station operable around both clock and calendar. Both stations were operable only on a "stand-by" basis until the work was mostly complete, about December of 1951. At both stations the spacing between array instruments was reduced from 3700–4400 feet to 3,000 feet to improve signal coherence. The rehabilitation also included adding Perkin-Elmer break-circuit amplifiers at these two stations—a recently developed state-of-the-art low frequency amplifier that permitted recording both high-and low-gain outputs simultaneously.

The Laramie Analysis Center

For the first few months after becoming operational, seismic data were sent to the Troy, N.Y. office of Beers and Heroy for analysis. Then, in September 1951, we began to establish a more permanent secure center for analyzing the seismic data from the growing station network, in rented space at the University of Wyoming (the center was later moved to commercial space on Grand Avenue in downtown Laramie, Wyoming). The center, to become known as the Laramie Analysis Center (LAC) was operated under my direction, with funding and oversight provided to B&H by AFOAT-1. Core staff members[6] transferred from the Troy, N.Y. office began to prepare for analysis of the data. In the interim, while capabilities were being developed in Laramie, Brad Leichliter and Bill Brooks conducted analysis at AFOAT-1 Headquarters in Washington, D.C. Additional staff members were recruited from students, and spouses of students or staff, of the University of Wyoming. They were first trained in the analysis of seismograms, and then initiated into an around-the-clock operational routine once the numerous problems of establishing and equipping a secure center were solved. The basic objective was to detect and locate the epicenters of potentially suspicious seismic events in and near the USSR and China, and to report them promptly to AFOAT-1.

Data from the seismic stations, arriving in two forms, became the basis of the analysis at the LAC. The initial data from most of the stations were transmitted over military circuits as SECRET level telegraphic messages containing standard alpha-numeric descriptions of each signal detected by analysts at each station. These reports included the arrival time, amplitude and period of each detected signal, or "phase," as well as the analysts' interpretation of the nature of the phases, e.g., P, S, etc. Each station was identified in each of its messages by a four-letter animal name, e.g., DEER, (Pole Mountain), BEAR, (Ankara), GOAT, (Fairbanks) that could be used as an unclassified reference as long as the station's location and function were not revealed. It also identified the message as seismic if it originated from a location where other types of sensors were also in operation. Such messages were prepared, encrypted, and dispatched every eight hours from each station. These messages were received, decrypted, and passed to the LAC staff by an Air Force team that supported the LAC with classified communication services.

Implicit in this process, an analyst at each station had to change the photographic film recordings at 8-hour intervals. The developed film recordings were then analyzed at the station, and each detected signal was reported telegraphically. Then the SECRET films were sent to the LAC by classified mail; these films formed the basis for the second form of analysis. Analysis was conducted on a Pentastrip Film Viewer (see Melton, 1981) that permitted up to five 35 mm film recordings to be viewed simultaneously, and manipulated to superimpose, or shift in relative time, to aid in the analysis of recorded signals. These could be 3-components of motion from a single station, array outputs from five stations, high-gain and low-gain recordings from the same instruments, or other combinations as needed.

Thus the analysis at LAC took place in two stages: first a prompt "preliminary analysis" based primarily on telegraphic reports, and then days later—paced by classified mail and diplomatic pouch schedules—a "final analysis" based primarily on reanalysis of film recordings. Once all signals had been detected, their times of arrival at each station measured, and the phase type identified, the main analytical tasks were "association" and "location." Association meant selecting all signals recorded by the various stations that had emanated from a given single seismic event. Location was the follow-on process of using arrival times of all associated signals to determine the coordinates and time of occurrence of the event.

Why Laramie?

Which brings us to the question of "Why was Laramie chosen as a place to carry out this function?" The answer is that the Wyoming "tripartite" network had been established nearby—three stations 120-200 km apart —and Laramie was the city most nearly central to the stations. These stations were built on massive, hard rock (granite), and, as expected, were exceptionally quiet except when winds were high or a Union Pacific Railroad train thundered by 10–12 miles away, as frequently happened at Pole Mountain (this station was closed and moved to Pinedale, Wyoming a few years later as a consequence). The relative times of arrival at the three stations of P waves from an event could be used to calculate the azimuth (direction) to the source. The arrival times also gave the apparent surface velocity across the tripartite network, which is a measure of the distance to the source—the greater the speed the greater the distance. With an azimuth and a distance we could estimate the latitude and longitude coordinates and the origin time for an initial, trial epicenter. This concept required that the tripartite be large enough to obtain a reasonably accurate epicenter, but small enough to contain good visual waveform correlation between stations. Good visual correlation was crucial, especially for estimating distance. Miscorrelating P waveforms by one cycle (say, one second) could cause errors of 300 to 800 km, depending on distance to the epicenter. The 100–200 km spacing for the tripartite had been selected from other data and had been found to be effective using signals from the Greenhouse nuclear tests.

The trial epicenters obtained from the Wyoming tripartite were the key to the association process. Using this information, an estimate could be made of the expected time of arrival of signals at other stations. If candidate signals had been reported or were found from film analysis near the time expected from the trial epicenter, then these signals could be assumed to be from the same event, and a second estimate could be made of the epicentral location and time of occurrence incorporating the additional arrival times. To the extent that the arrival times fit, confidence that the signals were correctly associated increased, as did the accuracy of the epicenter.

The Art of Epicentral Location

For obtaining the final epicentral coordinates, the state of the art at the time was to use a globe three or four feet in diameter. The globe was accurately ruled in geocentric latitude and longitude lines. The globe actually then in use by the USC & GS for "Preliminary Determination of Epicenters" had brass pins embedded at the locations of key stations that dependably and promptly reported data. It also had a calibrated tape, marked in units of great circle degrees, that could connect to the station pins. The seismologist made an estimate of the time of occurrence of the event under study, perhaps using the difference in arrival times of P and S at some particular station, then using the reported arrival time of P waves at each station, looked up the corresponding distance in an empirically derived travel-time table. Using his calibrated tape, he drew arcs on the globe representing each station's distance. If the arcs all intersected in a point, he had found a good location and origin time. If not, he adjusted his origin time and tried again. This was the "analog computer" of the day. Achievable accuracies were about ± 50 km (roughly one-half degree) in latitude and longitude.

A second method was mentioned by MacElwane and Sohon (1936): "Next to the use of a large, calibrated globe, the simplest way to find the approximate epicenter is to use one of the stereographic projection methods." This was, in fact, the method adopted for use at LAC (for more detail, see Appendix C).

We were focused almost entirely on an "Area of Interest" defined as the territories of the USSR and China plus a buffer region of about 50 km, corresponding to our routine epicentral accuracy. Events outside of this region were of no interest, and were analyzed only to the extent needed to explain all detected signals that otherwise might have originated from an event in the Area of Interest.

Geiger's least-squares method, using electric-driven, mechanical calculators, could be employed for greater accuracy on events of unusual importance. However, this method could take many hours of computation, and thus was rarely used. Professor Byerly instructed his graduate students, "If an important earthquake occurs, and you want to be forever cited as *the* authority on its epicenter, do a least-squares analysis. No one will ever repeat the calculation." The stereographic method remained in use until about 1960, when solid-state digital computers became available.

The work at the LAC developed into a routine. Data from the three stations of the Wyoming tripartite initiated the analysis. Film recordings were delivered by Air Force operators from the nearby Pole Mountain Station. All other stations provided data by telegraphic reports. For events in the Area of Interest of unusual characteristics, or when needed to resolve ambiguities, the LAC would request that the film from the Encampment and Douglas stations be delivered immediately by automobile. In that high mountainous country, there were numerous "adventures" experienced by the drivers, especially during winter. But, by-and-large, the process worked well. Because the AEDS stations were each at carefully selected quiet sites, the sensitivity of the network to low magnitude events in the USSR and China was unparalleled; and we soon redefined the seismicity of that area. Although we didn't know it at the time, data on seismic events in the Soviet Far East, released by the USSR ten years later, indicated that the AEDS seismic system was more sensitive than the Soviet standard network in that region.

Success!

A second and a third atomic explosion took place in the USSR in late September and mid-October of 1951, but they were of modest yield, the U.S. seismic system was still in an early developmental state, and the shots were not detected seismically.

The first real payoff came with the detonation of the fourth Soviet nuclear explosion, *Joe-4*, on 12 August 1953. This explosion was a 400 kt thermonuclear shot on a tower at the Semipalatinsk Test Site (STS). It was recorded in Wyoming, Alaska, Turkey and at a location that may still be classified. The recordings at these stations were excellent, and the geometrical arrangement of the stations with respect to the epicenter was good. For the first time, the U.S. had the data required for an accurate location of the STS. (The general region of the test site was known from acoustic signals from the second and third Soviet tests; but acoustic signals—affected by strong winds—are inherently less capable than seismic signals for determining locations.)

A follow-on to this event (and a slight historical digression) took place in 1956, several months after I had accepted a job with AFOAT-1 and moved to a suburb of Washington, D.C. I was ushered into a highly secure facility in the Washington D.C. area, where I briefed a group of U.S. military and civilian people on seismic detection of *Joe-4*. I was particularly grilled on the accuracy of the calculated location. My rather optimistic estimate at the time was that our seismic epicenter was probably within 10 km of the true location of the blast. End of story, or so I thought—I was ushered out, and given no real explanation for the briefing, or the identities of the audience.

Later, in late October and early November of 1962, the Soviets conducted three large high altitude and near-space tests from their Missile Testing Range near Kapustin Yar in the Astrakhan Region. Surprising as it may seem, these were detected seismically. We extracted as much information from the seismic data as possible, and subsequently I was named to a panel to evaluate the tests. There, another panel member introduced himself, and identified himself as part of the audience of my 1956 briefing on *Joe-4*. He went on to say that he was the pilot of the U-2 aircraft that first photographed the STS.[7] He made the flight, expecting that the seismic epicenter was wildly off the mark, but Lo! There was a large circular patch of burnt earth (fused into a black glass, as we learned much later) within 10 km of the seismic estimate!

THREE

Early 1950s to the *Rainier* Explosion

Even before deployment of the AEDS seismic stations and their successful use to detect *Joe-4* at long range, seismologists were exploiting smaller nuclear explosions detectable only at much shorter distances. Research proceeded along two tracks: academic seismologists used them to improve knowledge of the earth's crust and upper mantle for the most part, while AFOAT-1 scientists concentrated on such problems as how to differentiate explosions from earthquakes and how to determine an explosion's yield from seismic waves.

As will be outlined in this chapter it was not until far larger explosions were detonated later in the 1950's and the first underground test, *Rainier*, was conducted that numerous other long range detection questions could be seriously addressed.

Opening of the Nevada Test Site

The Nevada Test Site (NTS) became operational in late January of 1951, commencing with Operation Ranger. The practice throughout the U.S. nuclear testing program was to organize tests into an "Operation" consisting of one or a series of shots usually at one or two nearby locations. In the early operations it was customary to identify the individual shots by alphabetical names taken from the phonetic alphabet used by the U.S. military forces. As a result, there were four *"Ables," "Bakers"* and *"Easys,"* two or three *"Charlies"* and *"Dogs"* and so on. As a result, it became necessary to include the Operation name along with the shot name to differentiate events, e.g., "Greenhouse *George.*" This dual name convention persisted for a number of years, even after dropping the use of alphabetical names and choosing names from other categories, e.g., mountain names (*Rainier, Hood*), American Indian tribes (*Zuni, Cherokee*), tree varieties (*Aspen, Holly*).

Five air-dropped explosions were conducted during Operation Ranger between 27 January and 5 February 1951, ranging in yield from 1 to 22 kt, and detonated at elevations slightly above 1000 feet. Later, on 31 March 1951, about 60 tons of chemical explosives were detonated in a tunnel in a quarry near Corona, California. The Corona shot, as well as the larger shots at NTS, was recorded seismically out to distances of about 400 km, giving an early indication of the increased seismic coupling (fraction of the explosive energy converted into seismic waves) of underground explosions, relative to atmospheric shots. Gutenberg (1952) analyzed the travel-time data from both NTS and Corona, and used them to refine knowledge of P-wave velocities and crustal structure in southern California and Nevada. Aside from a comment that the shear waves can be expected to be relatively small in explosions, the essence of the paper was academic, having little to do with nuclear test detection, per se. At about this time, Dean Carder and Leslie Bailey of the U.S. Coast and Geodetic Survey began a systematic program of collecting and analyzing data from NTS shots, sponsored at least in part by AFOAT-1.

AFOAT-1 Seismic Experiments

Operations Buster and Jangle took place as an essentially continuous series of seven tests in late October through November of 1951. They included airdrops of 3.5 kt to 31 kt, detonated slightly above 1,100 feet; and also a pair of 1.2 kt explosions, one placed one meter above the surface and the other buried at a depth of 5 meters. The latter two tests had been proposed by the Department of Defense to investigate the effects and possible military value of such near-surface detonations. There were numerous other investigations, including projects sponsored by AFOAT-1 to study Long Range Detection. The surface and shallow underground explosions originally had been approved by President Truman for execution on Amchitka Island in the Aleutians (Pouton, *et. al.*, 1982), but that location had been reconsidered as the comprehensive test plan developed, when it became apparent that the Aleutians were unsuitable for many of the proposed research projects.

Under the title "Seismic Waves from A-bombs Detonated Over a Land Mass, Buster Project 7.5, Jangle Project 7.2," AFOAT-1 organized a large-scale seismic program consisting of three separate projects. Under the first project, the U.S. Coast and Geodetic Survey operated displacement meters, tiltmeters and accelerometers on the Nevada Test Site within 20 km of the explosions. They also analyzed data from permanent seismic stations at Pierce Ferry and Hoover Dam, Nevada. This work, under Dean Carder, attempted to determine the amount of seismic wave energy coupled into the ground from the several planned shots.

A second, much larger project under the technical direction of the Naval Ordnance Laboratory (NOL) studied the seismic waves at intermediate distances. Under this project 40 seismic stations were deployed, mainly at distances of 20 km to 450 km along a profile northwestward from NTS almost to Reno, Nevada, and along a similar profile southeastward almost to Prescott, Arizona (see Figure 1.1). The project was conducted jointly with the 1009th Special Weapons Squadron (the 1009th SWS was the operational arm of AFOAT-1, which functioned essentially as a Headquarters). The NOL supplied the seismographs and trained 60 airmen of the 1009th SWS to operate the equipment. Melton (1981) describes the "NOL seismometer" as a small, short-period (~1 second period) instrument designed to go down a borehole. It was developed under the direction of J.V. Atanasoff, (later recognized as the true inventor of the digital computer), and it had been deployed earlier to record the Helgoland explosion of 19 April 1947 (more on this explosion later). It employed a capacitance-bridge transducer, and an Esterline-Angus pen and ink paper strip recorder. When deployed at various locations around the Nevada Test Site, these seismometers were installed on outcropping bedrock, and "protected" by placing their packing crates over them—a common practice in seismic field operations of the day.

Results of interest from the first two projects were that the seismic amplitudes observed from the surface shot *Sugar,* and the shallow underground shot *Uncle* , were almost equal; which was perhaps not surprising in view of the small depth of burial. But perhaps unexpected was the observation that the 3.5 kt *Baker* shot at a height of 1,118 feet produced comparable amplitudes to both 1.2 kt near-surface shots, implying a reduction in seismic source effectiveness of only a factor of three at that height. The USC & GS estimates indicated that only a fraction of one percent of the bomb's energy went into seismic waves. Among the airbursts, all at nearly the same height, seismic amplitudes varied as the 0.6 to 0.75 power of the yield.

A third project, under Beers and Heroy, attempted to study seismic propagation to longer ranges, using data from the operational AEDS stations in Wyoming, and from temporary stations in Oklahoma, Alabama, and New York (see Figure 1.1). These stations had been expected to bracket a predicted "shadow zone," or zone in which P-waves were weaker than they were at both lesser and greater distances. Stations in Wyoming were expected to be near the edge of the shadow (911-1,069 km). An array station at Fort Sill, Oklahoma (1,610 km) was expected to be just beyond its minimum, and stations at Fort McClellan, Alabama (2,760 km) and Carmel, New York (3,640 km) well beyond the shadow (see the final paragraphs of this chapter for more explanation). The shallow buried shot and the air bursts were detected by the Wyoming arrays out to 1,069 km, but none of the shots were detected at more distant stations, leaving questions about the existence or extent of the shadow zone unresolved.

On our research for "discriminatory characteristics of blasts," I reported that it had been "hoped that the Greenhouse seismogram characteristic of a one-second pulse, followed after about 20 seconds by a short train of waves increasing in period to about 1.2 seconds"(Romney, 1952) would be found. Correlation analyses, however, revealed no such characteristic. Our studies also showed that the P waveforms recorded at two of the Wyoming stations were essentially uncorrelated, suggesting little hope of finding any distinct characteristic indicative of an explosion in such waves. This concentration on searching for an explosion "signature" in the P-waves reflected the fact that only P-waves were (and are) detectable at great distances from small explosions.

Development of the AEDS Seismic System

As nuclear testing continued at NTS and in the Pacific, the staff of the Laramie Analytic Center used the resulting seismic data to develop operational skills. We found that seismic signals from some of the larger NTS tests were detected at the Wyoming stations and in Alaska. (On occasion we also detected fallout from certain tests—photographic seismic recordings that we received from stations in Wyoming sometimes were speckled with black spots where radioactive particles had exposed the film.) Our main focus on the seismic data remained on trying to determine the "signature" of explosions (characteristics that would identify the event as an explosion and not an earthquake). But we also tried to determine the dependence of signal amplitude on yield, height of burst and distance from the explosion—both departures from the more classical work going on elsewhere.

The stations near Encampment and Douglas, Wyoming, continued operations by technicians of Beers and Heroy until 1952, when they were turned over to Air Force teams. AFOAT-1 rapidly built and incorporated other stations into its "Interim Operational System." By May 1952 a new station began operations in Korea, at a site that may still be classified. "Team 311," became operational in August 1952, as did one at Thule, Greenland in October, 1952 and one at Camp King, Germany, shortly afterward. Standard equipment at these sites consisted of four vertical component seismometers in a linear array with approximately one-half to one km spacing, plus two horizontal instruments. All were Benioff short-period instruments that recorded photographically on 35-mm film.

In view of the importance of accurate time, AFOAT-1 had contracted with American Time Products of New York to develop a suitable timing system. The resulting device was controlled by an electrically driven tuning fork in a temperature-controlled oven. To insure continuous timekeeping in the event of a power outage, a British-made synchronome pendulum clock driven by a separate battery served as a back up. The system delivered timing pulses, the duration of each pulse coded to identify ten seconds, one minute, five minutes, half-hours and hours. It also delivered 60 Hz power for driving the recorder motors at a constant speed. Its accuracy was considerably better than the chronometers or pendulum clocks that were standard in seismic observatories at the time. Corrections to Greenwich Mean Time were obtained through comparison of timing pulses with time signals broadcast from astronomical observatories in Washington, London, and other locations.

To avoid excessive electrical losses in the cables connecting the seismometers to the recording galvanometers, it was necessary to place the central recording station near the sites where the array sensors were located. In the case of the Encampment station, sensors and central recording station were physically located in Dead Horse Park, about 16 km west of Encampment and at an elevation of about 3,000 meters near the continental divide. Housing for the dozen or so operators was in Encampment. The rigors of commuting back and forth over primitive roads in those high mountains were severe. Bill Braukman, who was stationed there, has recently reminded me that Detachment 140 (the Air Force Team) was the only Air Force unit with a team of horses assigned to it because of the rugged terrain. In winter, tracked vehicles were required at each of the Wyoming stations, as were snowshoes.

Lightning was found to be a serious problem to the operation of the seismic arrays, particularly in Wyoming during warm weather. Massive thunderheads formed over the mountains when moist air from the Gulf of Mexico was driven westward up the slopes of the Rocky Mountains. The Spiral-4 cables connected to galvanometers that were sensitive to microvolts turned out to be antennas. Even though the cable was shielded, and conductor pairs were twisted, nearby lightning strikes induced sufficient currents to destroy the galvanometers, knocking the seismographs out of operation. Our engineering staffs became experts in the lightning protection measures of the day, but a solution remained elusive for years.

Seismometers in each array had been sited on granite outcroppings to minimize earth noise, but wind-induced noise was frequently a problem because the shelters built around the seismometer and its concrete pier did not give sufficient protection against the high winds that blow in Wyoming. To reduce the effect of wind, soil was mounded up around the shelters and rounded to reduce turbulent air flow. But "wind noise" continued to be a problem, yielding only slowly to measures such as placing seismometers in sealed steel tanks buried in shallow soil and bonded to bedrock. This, however, was possible only after remote calibration methods were developed and tested: the calibration procedure at that time was based on measuring the deflection caused by suddenly lifting a 1-gram weight off the 100 kg Benioff seismometer mass—a hand operation requiring weekly access directly to the seismometers.

A training school was established near the array site at Ft. Sill, Oklahoma in the spring of 1952. Billy Brooks, Braden Leichliter and I assisted Lt. William Fennell in writing training manuals and operating instructions, and in conducting training classes for the airmen assigned to the initial training course. For many of us who participated in that course it was our first experience with opaque dust storms, the sight and sound of tornadoes, and hail stones that dented automobiles and stripped neon tubing off display signs. It was also an opportunity to experience soft spring days among the bison and wildflowers in the beautiful Wichita Mountains Wildlife Refuge located nearby.

The development of a new amplifier suited to seismometers at quiet sites, and having better operating characteristics and greater power output than the troublesome Perkin-Elmer amplifiers, added new capabilities to the system in 1952. The amplifier continued to use a galvanometer to interface with the seismometer because of its low noise level, but the light beam fell on photoelectric tubes instead of photographic paper with the result that inputs of microvolts could be amplified to outputs of volts. This permitted recording the data 10 or more kilometers away from the instrument sites. It thereby solved the commuting problem at stations like Encampment, where access was difficult. This, in turn, permitted the central recording stations to be moved to a more accessible location.

Stations at Thule, Greenland, and Camp King, Germany proved to be too noisy, the former because of interference from "ice-quakes" as nearby glaciers lurched; and both sites were adversely affected by short period microseisms, apparently originating in nearby oceans. Both were closed down and the equipment used at other locations. New stations were installed at Sonseca, Spain, Alice Springs, Australia and elsewhere. For these new stations optimal separations for the seismometers were found by experiments using several seismometers placed different distances apart and in different directions. With our new amplifier we could use the output from one detector to drive the horizontal sweep of an oscilloscope, and the output from another to drive the vertical sweep. We could assess the coherence of the noise by photographing the resulting Lissajou figures for a minute or so. If the outputs were highly coherent, as would be the case at small detector separations, we recorded a nearly linear figure sloping at a 45° angle. Increasing the separation reduced coherence, until we finally recorded approximately random motion within a roughly circular region. Our array design criterion was to seek incoherent noise, but coherent signals; so we selected the separation for the sensors at the minimum distance where the noise was incoherent, relying on the longer wavelength of the P-wave signals to maintain coherence.

The phototube amplifier provided sufficient signal voltage to record signals at both high- and low-gain from the same detector, as well as to drive other electronic devices. Perhaps the most useful of these other devices was developed under the direction of Martin Gudzin of the Geotechnical Corporation for our use in transmitting and receiving seismic data over telephone lines. This device generated seven discrete tones that could be transmitted over a single telephone voice line. Each tone was frequency modulated by a seismic data channel; for example, the four vertical sensors of the array, two horizontal sensors, and the combined array output. Similar devices now in common use are called "modems." In the early 1950's Western Union, the U.S. telegraph monopoly, had the sole right to transmit data. The Bell Telephone Company had exclusive rights to transmit voices. Ultimately, Ben Melton and his associates in AFOAT-1 succeeded in breaking down these monopolies by arguing that only telephone lines were built to transmit faithfully the tones from the frequency modulated system. After some experimenting, in 1954 and 1955 seismic data was transmitted operationally by telephone lines to the Laramie Analysis Center from the Wyoming Tripartite stations, obviating the need for automobile trips from the stations to Laramie. Our system was called the "Zipagram system," and its use marks the beginning of legal data transmission over telephone lines in the U.S.

Britain Joins the Nuclear Club

The first British nuclear test took place on 3 October, 1952 at Monte Bello Island, a remote speck of land just off the northwestern tip of Australia. Next, two small tests were conducted in 1953 at Emu Field in South Central Australia, followed by two more at Monte Bello in 1956. But then, in 1956, four explosions were detonated at Maralinga, in south central Australia. Doyle (1957) took advantage of the nearly straight trans-Australian railway to place seismographs about 200 km apart west of Maralinga to obtain the first reliable data on the crustal and upper mantle seismic velocities and structures under Australia, a continent of extremely low natural seismic activity.

Powerful Seismic Sources and the Earth's Core

The thermonuclear era had been foretold by the Greenhouse *George* explosion, the world's first to contain thermonuclear components; but was firmly established during Operation Ivy with an experimental device yielding 10.4 Mt. This device, Ivy *Mike*, detonated at Enewetok on October 31, 1952, was well recorded seismically at numerous stations out to distances of more than 140°, although few seismologists realized that the signal recorded at their station came from the bomb. In those days yields and times of nuclear explosions were not announced, so to the seismologist at a single station, or small network of stations, the P-wave signal was just another one of the numerous similar small signals routinely recorded from earthquakes at unknown locations.

However, Gutenberg had adequate security clearance, and he knew the time and place of the bomb. Without explicitly identifying the source, or its coordinates and time, Gutenberg collected seismograms from stations around the world. The results of his analysis were promptly reported in the Proceedings of the National Academy of Science under what Professor Keith Bullen of the University of Sidney, Australia, was to call the "enigmatical title": "Travel Times of Longitudinal Waves from Surface Foci" (Gutenberg, 1953). Bikini *Baker*, as a source of seismic waves detectable at large distances, was well known at the time, but other large sources were unknown to most scientists, so the use of "foci" was truly enigmatical.

In this paper Gutenberg explained only that "travel times of seismic waves propagated through the earth are based on earthquake records" and that it is "highly desirable to compare them with experiments where latitude, longitude, depth and origin times are accurately known from independent sources." He reported significant discrepancies from his previous travel-time tables and from the Jeffreys-Bullen travel-time tables, and he completely revised and published a table for P-wave times from a surface focus. He also reported five recordings of PKIKP (longitudinal waves through the earth's outer and inner core, see Figure 3.1) at distances between 110° and 140°, and he noted that three of these recordings showed precursors 5-15 seconds before the main pulse. He concluded that these anomalous precursors were probably due to diffraction. Burke-Gaffney and Bullen were to reach the same conclusion the next year.[8]

Operation Castle, two years later in 1954, consisted of five large explosions at Bikini and one at Enewetok. They were all detonated on the surface, either on islands of the atolls or on barges, and they ranged in size from 110 kt to 15 Mt. Burke-Gaffney and Bullen (1957) noted an impulse on the Riverview, Australia, seismograms that was consistent in time with press reports on the 15 Mt *Bravo* explosion of March 1. (These reports to the press, a departure from the normal practice, resulted from an excessive amount of radioactivity injected into the air.) They also found related signals on that date reported in bulletins mailed to them from other seismic stations. During the next three months, three other multi-megaton explosions took place, and these also produced signals reported in many bulletins. Burke-Gaffney and Bullen noted that none of the bulletins had associated their reported signals with any of the explosions; they were simply reported in the same way as signals from any small earthquake, once again suggesting that discriminating between earthquakes and explosions might be difficult.

Bullen felt it necessary to raise the question of the ethics of using data from explosions of another country, even in pursuit of science. He posed the question in April 1956 at a meeting of the Council of the International Union of Geodesy and Geophysics. Dr. J. W. Joyce, a U.S. representative at the meeting, undertook to obtain an answer in Washington, D.C. He subsequently reported back that the United States had no objections of any sort to the publication of information obtained in the course of regular scientific work. Burke-Gaffney and Bullen proceeded with their analysis.

Data reports on the four shots were available to them from 18 stations at distances ranging from 33° to about 144°. They were able to confirm the corrections to the P-wave travel-time curves, as reported by Gutenberg from the Bikini *Baker* shot, using an assumption that the times of the shots were exactly on the minute, as had been the case for the *Baker* underwater explosion. They also found that the travel times of P-waves to stations in Australia and in the U.S. at the same distances agreed within one second, giving good evidence that the Earth's interior was quite uniform from region to region. Their most notable result, however, was that they observed weak precursors about 7-11 seconds before a stronger impulse at Pretoria and Kimberley in South Africa, and at Tamanrasset in southern Algeria, at about 137° to 141° from Bikini—probably the same three stations where Gutenberg had noticed precursors in 1953.

These waves must have penetrated into the Earth's core, which intercepts waves that have traveled in the mantle to a distance of about 103°. Burke-Gaffney and Bullen believed the waves might shed light on the nature of the core. The existence of the Earth's core, into which these distant waves had penetrated, had been adduced near the end of the 19th and beginning of the 20th centuries. The two most important early clues came from measurement of the Earth's density and tides. Just as the sun and moon cause tides in the oceans, they also cause smaller tides in the Earth itself. Earth-tides, moreover, were found to be substantially larger than they would be if the earth were solid rock throughout. This led to the hypothesis that a major portion of Earth's interior must be liquid. The average density of the earth was found to be much greater than rocks found at the surface, leading to the hypothesis that the deep interior was made of some much denser material, probably a metal. Arguing from the existence of both iron- and stoney-meteorites it was speculated that the metal most likely was predominantly iron, since meteorites were believed to be fragments of celestial bodies destroyed by a cataclysm of some kind. The rapidly developing science of seismology provided both data and understanding of seismic wave propagation that Gutenberg had used in 1914 to measure the radius of the core (3,500 km, slightly larger than Mars). He also found that it had a sharp boundary, where the longitudinal wave velocity dropped suddenly from 13.5 to 8.5 km/second as the waves entered the core. By 1926 the great English mathematician and physicist Sir Harold Jeffreys had proven conclusively that the core was liquid, and the transition from solid to liquid caused the velocity change.

Harold Jeffreys was another giant figure of the era. Whereas Gutenberg was unexcelled at observational seismology, Jeffreys was the premier theoretician in seismology; and indeed, was an eminent theoretician in numerous other fields of physical science. He was the author of landmark papers on the origin of the solar system, on tides, on meteorology, and on the internal constitution of the Earth and other planets. His work in collaboration with Keith Bullen on the times of travel of P and S waves within and through the Earth resulted in the "J-B Tables," a standard for seismologists from 1939 throughout much of the remainder of the twentieth century. His monumental book, *The Earth*, described much of what was known about physical aspects of its subject, and gave the mathematical underpinnings of each subject.

While Jeffreys applied mathematical rigor to the science of seismology, he well understood that the Earth is far too complex to describe accurately by analytical models, and that theory, at best, only roughly described the Earth. Perry Byerly told of "punting" on the Cam River with Jeffreys while on sabbatical at Cambridge. On this relaxed Sunday outing, Sir Harold remarked, "You know Perry, to a first approximation the punt is as wide as the Cam." And in geophysics generally, much of what is "known" about the Earth remains in the realm of approximation, to some degree.

P-waves that entered the core (called PKP—refer to Figure 3.1) had been shown to emerge at distances beyond about 142°-145°, a distance shown by travel time data and amplitudes to mark a caustic (focus). However, other late-arriving P-waves (apparently delayed by traveling within the lower velocity core) had been observed at significantly smaller distances than 145°, and these were not readily explained if the core was uniform throughout. At the time, there were two schools of thought about these waves. One explanation was that they were diffracted waves from the caustic at 142°-145°. However, Dr. Inge Lehmann had pointed out in 1936 that the observed waves were probably too strong to be diffracted. She proposed the existence of an inner core, marked by an increase in P-wave velocity, as a second alternative to explain them.

At the time, "Miss" Lehmann, as she was usually referred to by colleagues, was Chief of the Seismological Department of the Danish Geodetic Institute. Her seismic stations were located at just the right distance to study waves through the core from energetic earthquakes in the South Pacific, and so it was natural that she should be interested. Her great discovery about the inner core was questioned by some seismologists for many years to come, but her insights were true. Miss Lehmann became the internationally known and respected "Grand Lady" of seismology. Few women achieved top ranking in science in the first half of the twentieth century, and her attainment of such ranking required strength of character as well as great scientific insight. She was a frequent visitor to the United States, continuing her research after retiring from her position in Denmark; her last scientific paper was published in 1987 at age 99 (see the "Memorial Essay" by Bolt and Hortennberg (1994)).

Gutenberg and Richter (1938) provided travel-time evidence supporting Miss Lehman's proposal, as did Harold Jeffreys in 1939. However, Gutenberg and Richter proposed a transition zone between outer and inner core, in which the velocity gradually increased. They noted that no reflected waves had been found, as would be expected if the boundary was sharp. After obtaining copies of the *Castle* seismograms from the three stations, Bullen and Burke-Gaffney (1958) presented good evidence that the weak precursors observed from Operation Castle's shots were, in fact, diffracted from the caustic, and thus the strong waves that followed must have traveled through an inner core, confirming Miss Lehmann's interpretation—today almost universally accepted. Bullen and others later showed that, in all probability, the inner core is solid.

That same year, Gutenberg (1958) gave yet another alternative explanation for the observed precursors. Although he had adopted the idea of an inner core years before, and continued to support the concept, he now did not agree that the precursors were diffracted waves. He suggested again, as he had in 1938, a transition zone between inner and outer core where the core material was near its melting point and the velocity increased gradually with depth. He also cited studies predicting that dispersion would occur under this condition, in which short period waves (like the precursors) would travel faster than the somewhat longer period waves in the stronger pulse that followed—a significant shift from his conclusion of 1953.

As an aside, more than a decade later unequivocal evidence of the nature of the inner core's boundary also came from nuclear explosions. Evidence derived from the Large Aperture Seismic Array (LASA) in Montana (see Appendix E) was published by Engdahl, Flinn and Romney (1970). We had detected clear echoes, designated as PKiKP (see Figure 3.1), reflected at steep angles off the earth's inner core from nuclear explosions in Nevada (and from earthquakes in Central America and Alaska as well). These observations of sharp pulses of one second period established that the surface of the inner core is also sharply defined, as Bullen and Burke-Gaffney had thought, and that its radius is very close to 1216 km, as estimated by Bolt (1964).

However, to add a second historical note, the first convincing observations of PKiKP steep reflections were actually made in 1966-1967 at the classified AEDS seismic array station in Turkey[1]. Ronald Cook, then an Air Force seismic analyst at the station, noted that P-waves from underground nuclear tests at the USSR's Semipalatinsk Test Site were sometimes followed by a second, smaller P-type phase ten minutes later. He reported this, along with a question as to its nature to the seismic analysis center of the AEDS, where it was determined in all likelihood to be PKiKP. Calculations had predicted that this phase would be weak, and there was some doubt that it would be detectable. Knowing that the phase was detectable by seismic arrays designed to emphasize steeply arriving signals, and at a distance where it was relatively strong, greatly simplified our search for PKiKP using unclassified LASA data. Unfortunately, Ron Cook could not be given credit for his earlier discovery in the 1970 paper for security reasons.

Wigwam

On 14 May, 1955 a 30 kt device called *Wigwam* was detonated deep within the ocean about 1000 km west of Baja California, Mexico. Its energy was strongly coupled into the water and into the earth, producing a shock equivalent to an earthquake of about m_b = 6 - 6 1/2. This was about two orders of magnitude greater than that produced by an explosion of the same yield on the surface of the earth. It was felt by sailors aboard ships in the Pacific Ocean west of San Francisco, who were alarmed enough to make radio inquiries to a national news service as to whether a great earthquake had occurred. These inquiries were, in turn, passed to Professor Perry Byerly on that Saturday afternoon.

Byerly soon arrived at the Berkeley Seismographic Station, where I had been watching a most remarkable sequence of seismic waves being recorded on a pen-and-ink recorder (Figure 3.2). (I had returned to Berkeley in 1954 to complete my Ph.D. work, but retained my security clearance and knew of the plans for *Wigwam*). The seismogram began with a P wavetrain consisting of almost sinusoidal oscillations of 1.4-second period, followed after about 100 sec with an even larger and equally sinusoidal S wavetrain of the same period. Surface waves having a period of about 4 seconds followed S, interrupted after a few minutes by a very strong burst of high frequency waves. The latter was obviously the "T-phase" — sound waves that had traveled most of the distance to Berkeley in the ocean until they entered the crust along the steep coast — doubtless what the sailors had felt. I don't remember exactly what Byerly reported back to the press; Operation *Wigwam* was secret at the time; but he no doubt confirmed that a seismic disturbance had occurred under the Pacific, giving little other information.

Oliver and Ewing (1958) later reported on their analysis of *Wigwam* data from seismic stations on both east and west coasts of the U.S. They concluded that the short period surface waves that they observed were of two types not previously observed from Pacific sources, i.e., that the waves represented the short period branch of the Rayleigh waves and the first shear mode. The sinusoidal and persistent P and S wavetrains, such as those seen at Berkeley, they attributed to standing waves in the water (multiple reflections between the ocean's surface and bottom) leaking energy gradually into the Earth. Scientifically interesting as these results were, *Wigwam* had another impact: the signals were found to be easily detectable at great distances, and from their appearance at some stations they could not possibly be attributed to an earthquake. Furthermore, the intensity of the sound waves in the ocean had been enormous, as demonstrated by the sailor's reports and the size of the T-phase recorded at Berkeley and elsewhere. From that time on, we considered that detection and identification of nuclear explosions in the deep, open ocean should not be a problem.

Announcement of the First Underground Nuclear Explosion

At meetings of the International Association of Seismology and Physics of the Earth's Interior, held in Toronto, Canada, in early September of 1957, retiring president Keith Bullen chose "Seismology in our Atomic Age" as the topic for his address (Bullen, 1958). Earlier, and no doubt enthused by his recent successful exploitation of data from Operation Castle, Bullen had proposed to the presidents of the Royal Society of London and the Academies of Science of the U.S. and U.S.S.R. that nuclear explosions for the benefit of seismology be included in the forthcoming International Geophysical Year (IGY) in 1957-58. His proposal was not accepted for inclusion in the IGY for a variety of reasons (most cogently, poorly understood safety considerations), but now he had the opportunity to present his ideas to the scientists themselves, many of whom supported his proposals. (An excellent account of Bullen's and a number of their scientists' proposals for the use of explosions for science may be found in Barth, (2000)).

Coincidentally, the U.S. Atomic Energy Commission (AEC) announced, at that meeting, plans to detonate the world's first deep underground atomic explosion, to take place in Nevada in mid-September. Details of the yield and location were provided. This news generated considerable interest—even excitement—among some of the seismologists at the Toronto meetings; which led to a special, informal evening meeting to discuss the forthcoming event. The meeting was well attended by representatives of the U.S., U.K., U.S.S.R. and numerous other countries, noticeably not including the Japanese, who shunned all things nuclear at the time. P. L. Willmore of the U.K. and Canada played a prominent role in this meeting, and applied his experience with recording data from large explosions to guide the discussions. Pat Willmore had been the coordinator of the joint U.K./U.S. program to investigate seismic waves from the Helgoland explosion on 14 April 1947. Helgoland is a small island in the North Sea, heavily fortified during World War II. About 2.7 kt of munitions left over from the war in underground bunkers and chambers were detonated. More recently, Pat had conducted seismic measurements of the Ripple Rock explosion, in which a hazard to navigation near Vancouver, British Columbia, had been blown up with 1.3 kt of explosives.

The focus at the Toronto meeting was on how to make the nuclear explosion as useful as possible to seismologists. This included requesting the AEC to provide additional details and the precise time of occurrence after the fact for unclassified seismological research. There was also discussion of how to motivate geophysical prospecting crews to make observations of the explosions. This, in turn, might require that radio announcements be made of any last minute changes in detonation time because prospecting crews were equipped to record only seconds-long segments of data. The AEC offered, instead, that if any postponement was needed the shot would be delayed exactly 24 hours and announcements would be made to the press. There seems to be no record of this informal meeting, but the seismologists' views were heard by the AEC and all details needed by seismologists were made public.

In my first few years of work on nuclear test detection the possibility of a fully contained underground explosion had been mentioned from time to time, primarily as a testing method that would conceal the nuclear character of the test. However, when I mentioned a possible program of underground tests by the USSR as a reason justifying the U.S. AEDS seismic system—perhaps before the Boner Panel—the idea was dismissed as impractical. Bomb tests needed to take place in the air so that fireball measurements, radioactive end products and other diagnostics could be obtained, it was said, and tunneling to safe containment depths would be too expensive.

Nevertheless, Griggs and Teller (1956) proposed such a test, arguing that costs were comparable to those of constructing towers. They also cited the advantages of the freedom to test without delays caused by wind and weather, as well as the elimination of fallout. The proposal advanced slowly through the AEC, where concerns were raised about contamination of ground water and possible triggering of earthquakes. Perry Byerly was appointed as chairman of a scientific panel to look into those questions for the AEC. In a brief report to the AEC, Byerly's Panel concluded that there was no significant risk of either. Roland Beers, my former boss at Beers and Heroy, and a spokesman for the Byerly Committee, was credited with a substantial role in calming safety fears. As reported to me at the time, in his appearance as an expert witness before the Commission he was asked the size of the largest underground explosion that could be safely conducted at NTS. Beers calmly replied "about a megaton"—far larger than the planned test.

The *Rainier* Explosion

The underground explosion, a 1.7 kt device called *"Rainier"* took place a fraction of a second before 1700 GMT on September 19, 1957. The shot chamber was in a thick layer of tuff (a rock formed from consolidated volcanic ash) approximately 900 feet beneath the surface of a mesa on the Nevada Test Site. An initial report on the event, based on data from a few seismic stations, was issued by the USC&GS giving a magnitude estimate of 4.6. Dean Carder and Leslie Bailey of the USC&GS immediately requested seismograms from the stations in the United States, Canada and Alaska considered most likely to detect *Rainier*, based on prior experience in detecting earlier atmospheric explosions at NTS—some of which had produced larger signals than *Rainier*. As previously mentioned, Carder and Bailey were assigned to AFOAT-1 on a part-time basis, where I was employed by that time. And because of other commitments Carder had made, I took over his role in interpreting the *Rainier* data.

P-waves from *Rainier* were detected fairly consistently out to about 500 km, and at a few stations beyond. Reasonably good P-wave signals were recorded at Tucson, Arizona, (734 km southeast) and at Laramie, Wyoming, (1023 km northeast). Beyond Laramie, we detected exceedingly weak signals at Fayetteville, Arkansas (1964 km) and at College, near Fairbanks, Alaska, (3717 km). At both of these stations signals were detectable chiefly because signals from earlier, larger explosions at NTS had been detected, so we knew the precise time at which to look. It may be instructive to look at Figure 3.3, which shows the signal at Fairbanks. The point of the arrow marks the precise beginning of the signal; earlier motion (to the left) is micro seismic noise. Noting the prevalence of similar pulses on this portion of the seismogram, it can be seen that the signal would likely have been overlooked without prior knowledge of its existence and time of arrival. Figure 3.3 became the most widely seen seismogram of its time! It was published in numerous newspapers and news magazines, in part as "proof" that the AEC had lied to the public when it released a poorly-researched news item reporting that *Rainier* had occurred and had been recorded only to distances of several hundred kilometers.

In our discussion of the data (Bailey and Romney, 1958), we noted that the seismograms were very similar to those from any small earthquake. A portion of our unpublished paper reads:

Tinemaha...is due west of the shot, within a degree or two. It is interesting to note the large SH [transverse horizontal component of shear waves] component of motion on the N-S seismogram. We do not understand how this was generated by a spherical source. The radial component of SV [this is the component of shear waves in the plane of propagation] recorded on the E-W seismogram, arrived at the station slightly earlier than SH. The first P motion, although barely larger than the microseisms, is compressional, followed by a strong rarefaction; at a slightly greater distance it seems probable that the first motion might appear to be a rarefaction. The records are similar to those from "local" earthquakes.

Once again, these observations foreshadowed problems in differentiating between earthquakes and explosions.

The "local magnitude" as determined by averaging data from seven stations of the California seismic network was estimated to be $M_L = 4 \ 1/4$. Applying Gutenberg's (1956) newly revised relationship between magnitude and energy,[9] the energy in the seismic waves was about 3×10^{17} ergs, or about 1/200 of the total energy released by the explosion — a quite reasonable fraction according to the thinking at the time.

Rainier and the Shadow Zone

Classified stations in the Wyoming Tripartite network (840 to 1070 km) and in Alaska (3,700 km) detected clear signals from *Rainier*, and weak signals were detected at Fort Sill, Oklahoma (1610 km). A station at a now-forgotten location 280 km away also recorded *Rainier* using AEDS-type calibrated Benioff seismographs. These six measurements, obtained from the only stations where we were confident that P-wave vertical motions were accurately measured from well-calibrated instruments, were fitted to a curve of amplitude versus distance based largely on earthquake data, (but supported by data from large explosions in the Pacific and at NTS), to give an estimate of the amplitude versus distance for P-waves from a 1.7 kt underground nuclear explosion (the curve shown in Figure 5.5, adjusted to the amplitudes listed in Table 5.1).

The general shape of the curve was based on the work of Beno Gutenberg (1945) in his extension of the earthquake magnitude scale originated by Charles Richter in 1935. Gutenberg's corrections imply that the amplitude of P decreases rapidly between about 200 and 1000 km, and remains low until about 1500 km. Between about 2000 and 3000 km the amplitudes are as large as they were at 600 km. Thus, according to Gutenberg, there is a "shadow zone" between about 700 km and about 1500 km, in which P-waves are relatively weak.

Gutenberg (1948) later explained this behavior in terms of the velocity structure of the upper few hundred kilometers of the earth's crust and upper mantle. The concept is presented schematically in figure 3.4, updated somewhat in the light of later work. The principal structural elements are first, a low velocity crust, underlain by a higher velocity mantle. P-waves guided along the base of the crust and uppermost mantle (commonly called Pn-waves) die out rapidly with distance. Next, below the crust velocities decrease slightly with depth, reaching a minimum, which Gutenberg placed at various depths at different times, but typically at about 80-100 km. Decreasing velocities in the uppermost mantle bend waves entering the region downward, where they continue downward until the velocity increases once again, sending the waves upward to the surface, but at a distance beyond about 1500 km. The effect is to create a "skip zone," where only the weak crustal-guided waves exist.

The concept of a so-called "regional shadow zone" dates back to 1926, but was perhaps most clearly articulated and explained in the referenced paper (Gutenberg, 1948). It was not universally accepted, however. Along with changes in amplitude, physical considerations impose constraints on the times of travel of P-waves. Although Gutenberg, using data from western North American earthquakes, had reported travel times consistent with his amplitudes, Harold Jeffreys (1952), in a study of Mediterranean, Japanese, and Californian earthquakes had not. He concluded that "Gutenberg's discussion of the times is open to criticism, and the question of generality remains open." Tatel and Tuve (1955), in a study of seismic waves from blasts in the Tennessee region also reported no evidence of a shadow zone. Jeffreys' "question of generality" was, indeed, an open one as the work of Tatel and Tuve demonstrated, but it would take a decade to clarify the matter.

FOUR

Controlling the Atom
The Path to Negotiation

As outlined in the preceding chapters, seismologists took advantage of early nuclear tests—some even advocated more tests specifically for science—and used them as exceptionally useful sources of seismic waves to enhance understanding in the field. At the same time a public and political movement was underway to halt weapons development and testing. Ultimately, this movement led to formal East-West negotiations toward a ban on nuclear testing, as described in the following.

Postwar Proposals to Control Atomic Energy

Following the development of the A-bomb, and its use in Japan to end World War II, the U.S. sought to prevent proliferation of such weapons through international control of atomic energy. Banning atomic bombs before they had spread to numerous countries was the driving objective. Thoughtful people were appalled at the nature and extent of the damage done in Hiroshima and Nagasaki as the photographs flooded through the newspapers and newsreels after World War II, and a future nuclear war became a horror to be avoided at any cost.

The approach to this concern by Canada, the United Kingdom, and the United States—the three nations that had collaborated to create the bomb—was to offer to give up their monopoly on atomic technology and destroy all existing weapons in exchange for controls that would channel further developments exclusively into peaceful uses of atomic energy. Proposals to this effect were made within a few months after the end of World War II by the heads of government of the three nations in a joint declaration. The declaration also called for the creation of a commission within the United Nations to deal with atomic energy matters. A commission was, in fact, established by the U.N. General Assembly in January 1946.

In June 1946, specific proposals to share and control atomic energy were introduced at the United Nations by Bernard Baruch, President Truman's representative to the U.N. Atomic Energy Commission. Over the next two years, the Commission developed a comprehensive plan for the control of all aspects of atomic energy from the mining of radioactive materials to their end uses. However, in its second report to the U.N. Security Council on May 17, 1948, the commission stated that it had reached an impasse over its proposed safeguards which included the right of the U.N. Authority to inspect within the borders of any nation mining or possessing uranium or thorium. Although the majority believed that strong safeguards were essential to prevent misuses, the Soviet Union had rejected the plan "on the ground that such a plan constituted an unwarranted infringement of national sovereignty." We would hear this and other arguments from the Soviets against any form of inspection for many years to come. Subsequently, in November, 1948, the U.N. General Assembly approved the plan. Nevertheless, the Soviet Union continued to reject it, in part for reasons soon revealed.

Debate over the H-Bomb

The detection of *Joe-1* in 1949, years before it was generally expected, further fueled a burning debate within U.S. military, scientific and political circles on what to do about development of the much more powerful "super," or "Hydrogen-bomb," as the successful version later became known. The possibility that a thermonuclear fusion reaction between hydrogen atoms could be triggered by heat from the explosive fissioning of uranium -235 had been discussed in the early 1940's. Development of a weapon based on the fusion concept had been considered in the early days of the Manhattan Project, but was given low priority in the push for the fission weapons successfully built. To some, it now seemed a matter of urgency to proceed with this next logical step in weapons development; to others, it seemed a logical place to try once again, to convince the USSR to join us in capping further development of atomic weapons.

In the absence of clear national policy on nuclear weapons, scientists joined the public debate on this question. So great was their prestige stemming from scientific triumphs during the recent war, that their voices even dominated the debate. However, many of the issues were far from the domain of scientific expertise, having major military, cost and international political implications, among others. The result was a great split between scientists of opposing views, some of whom believed development was essential, while others were strongly opposed. This is not to suggest that scientists should not have viewpoints on such matters—far from it—but only to point out that the scientific method, by which agreement is ultimately reached throughout the scientific community did not apply here.

The preeminent scientific voices on these two sides of the question were Edward Teller, Director of the University of California's Radiation Laboratory in Livermore, California and J. Robert Oppenheimer, at that time chairman of the Atomic Energy Commission's General Advisory Committee and the director of the prestigious Institute for Advanced Study at Princeton. Teller had originated some of the early concepts for the "super," and he had a firm conviction that the USSR would proceed with the development of thermonuclear weapons without regard to U.S. actions. Oppenheimer spoke eloquently against its development, believing that U.S. restraint was essential to achieving international controls. The schism was deep, complex, and involved differing ethical judgments in addition to factors already mentioned; the reader is referred to Lapp (1956), Gilpin (1962), York (1989), and Bird and Sherwin (2005) for more thorough discussion and analysis of the events and views of scientists and politicians at that time. The consequences, however, were that the debate delayed the development of the American H-bomb—but not the Soviet Union's. President Truman's decision early in 1950 authorized thermonuclear development to accelerate, but without the urgency that some advocates wanted.

Early H-Bomb Tests

Testing of devices with thermonuclear components by the U.S. had begun with the *George* shot of Operation Greenhouse on 8 May 1951. This large fission device contained a small amount of a deuterium-tritium mixture (the double and triple weight isotopes of hydrogen). The explosion successfully ignited the mixture, producing the first thermonuclear reaction on earth. The significance of this explosion for the seismic system of the AEDS has been noted previously. *George* was followed the next year on 31 October with the *Mike* explosion of Operation Ivy, a 10 Mt *experiment* at Enewetok; the enormous device used liquid deuterium as its thermonuclear fuel. This, in turn, required large refrigerators to maintain the cryogenic temperature needed to liquefy the gas and the experiment was extensively instrumented to measure all details of the expected reaction. This 65-ton assemblage was, therefore, undeliverable, and could not be made into a weapon. However, as an experiment it was highly successful, and has been cited as the first of the true hydrogen explosions.

The next move turned out to be one made by the USSR. Joseph Stalin had died early in 1953, so he did not live to see the conclusion of the project he initiated in 1948 (Sakharov, 1990, pg. 94). However, his successor, Georgi Malenkov, announced on August 8, 1953, "The U.S. no longer has a monopoly on the manufacture of the hydrogen bomb." This was followed on August 12, 1953, by the detonation of the 400 kt *Joe-4* at the Semipalatinsk Test Site. It was a shocker to much of the scientific-military community of the U.S.! The Soviet confidence had been so high that the bomb was preannounced! Furthermore, U.S. analyses showed that the device was indeed thermonuclear, and it could be weaponized. As York (1989) explained, *Joe 4* was not a true member of the ultimate family of hydrogen bombs, as we know them today. Indeed, it took the Soviets more than two years to develop and test such a device. Nevertheless, it was an impressive accomplishment after only three fission tests, and only four years after their first atomic test.

Radioactive Contamination

Finally, on 1 March 1954, the *Bravo* shot of Operation Castle was detonated on the Bikini atoll of the Marshall Islands. It was the 44th U.S. test, after almost nine years of testing. It could be weaponized. Its yield was larger than expected: 15 Mt. As a consequence of its unexpected yield, fallout was much greater than planned for, forcing several hundred Marshalese and a number of Americans to be evacuated from the islands. Even worse, a Japanese fishing vessel, the "Lucky Dragon" cruising about 100 miles away received a heavy dose of radioactivity as a result of fallout. Not understanding the nature of the white ash that rained upon them, the fishermen allowed it to remain on the deck as they started for home. By the time they reached Tokyo, twenty-three members of the crew were seriously ill with radiation sickness. Fish contaminated by radioactivity were found in the Tokyo marketplace, and from other vessels as well. Although fallout had always been a safety consideration, and a reason cited for banning tests, this was the first major incident from nuclear testing. It seriously strained Japanese-American relations, as might be expected in view of U.S. attacks on Hiroshima and Nagasaki during World War II.

The *Bravo* incident incited world-wide concerns, as well as a call for suspension of all testing by some world leaders. Prime Minister Jawaharlal Nehru of India became a particularly vocal and persistent critic of atomic testing for the next several years. The following year, on 22 November 1955, the USSR conducted a 1.6 Mt test at their Semipalatinsk Test Site, an airdropped prototype of their future "true" hydrogen weapons. It resulted in extensive radioactive contamination of the test site and regions along the path of the wind-bourne debris, including Japan. Once again, international attention was focused on fallout, increasing the clamor for a test ban.

Arms Control Attempts

An even more significant force leading toward a test ban was the growing concern among world leaders about the pace and direction of the arms race (see, for example, Appleby, 1987) that had developed in the decade after the end of World War II. Initially, the U.S. and its Western allies had disbanded their military forces at a rapid pace. But faced with Soviet expansion into Eastern Europe in the late 1940's, its massive buildup of its conventional military forces, its blockade of Berlin for more than a year, and the invasion of South Korea in 1950 by the USSR's satellite state, the U.S. Government began rethinking its strategy.

Concerns were voiced clearly by newly elected President Eisenhower in his first foreign policy address, as he cited the need for alternatives to nuclear war on the one hand, or "a life of perpetual fear and tension," on the other hand. The public was also concerned; I remember storing drinking water and first-aid supplies in our basement in the 1950's. One neighbor actually constructed a fallout shelter in his backyard. Internationally known figures such as Bertrand Russell and Albert Einstein warned of the dire consequences of thermonuclear war, and appealed to governments to settle policy differences peacefully. Linus Pauling and others predicted tens of thousands of cases of leukemia would result from fallout, as well as uncounted birth defects among babies yet to be born. But the concerns among world leaders were broader than nuclear issues.

With growing unease about the "Cold War" and its attendant arms race, more emphasis was placed on the negotiating table, where there was discussion of many issues, characterized by the disarmament community's buzz-words of the day, including:
- "General and complete disarmament"
- "Surprise attack" and concepts to reduce its threat like "Inspection Zones" and Eisenhower's "Open Skies" proposal
- "Atoms for Peace"
- "Nuclear Test Ban," and its companion "Production Cutoff" (of nuclear fuels)

These issues were frequently linked together by one side or the other, and little progress was made in the negotiations. The U.S. position in the mid-1950's linked the nuclear test ban to a production cut-off. It could be argued that a test ban alone had little to do with reducing the arms race. However, it was known that verifying compliance with a production cut-off would require intrusive inspections of highly sensitive facilities, with no assurance that other hidden facilities were not still in operation. Test ban monitoring, on the other hand, seemed less intrusive, perhaps even easy, to people who understood the power of atomic bombs, but had little understanding of detection limitations. A treaty banning tests was seen to be a good first step in paving the way toward solving more threatening and difficult issues, even though U.S. policy-makers soon learned that verifying compliance would require more extensive monitoring facilities than those existing in the Atomic Energy Detection System. An international test ban treaty-monitoring organization had been proposed by the U.S. for operating monitoring facilities, although the idea had been rejected by the Soviet Union.

This had become a familiar pattern: the U.S. advocated disarmament measures accompanied by safeguards to insure compliance. The USSR embraced the same disarmament measures, but rejected the safeguards. On the test ban issue, Nicolai Bulganin, the Soviet Premier who had succeeded Malenkov, wrote a letter to President Eisenhower on September 11, 1956 containing the following:

> It is a known fact that the discontinuance of such tests does not require any international control agreements, for the present state of science and engineering makes it possible to detect *any explosion* of an atomic or hydrogen bomb *wherever* it may be set off. [Italics added.]

Clearly, this reflected the view of a closed society looking at open societies. The USSR felt no need for special measures to monitor testing by the western powers: they could simply read our newspapers. On the other hand, secrecy was deeply ingrained in the Soviet character as a vital element of their national security, and thus the U.S. would require highly sophisticated technical monitoring means to gain even approximate parity in ability to monitor the USSR.

Proposal for Technical Experts Meeting

A partial breakthrough came in mid-summer during the 1957 "London Conference" of the UN Disarmament Subcommittee. From our standpoint, the most important developments from the conference were that the Soviets agreed in principle to some form of inspection on their territory, and the U.S. agreed to consider the nuclear test ban separately from a cut-off of nuclear materials production.

Western representatives on the Disarmament Subcommittee promptly proposed that a group of technical experts be appointed by the Subcommittee members to design an inspection system to verify suspension of testing. This proposal stagnated, but the idea of a meeting of experts continued to appear in letters from President Eisenhower to the next Soviet Premier, Nikolai Bulganin, during the next several months.

Early Ideas on Seismic Discrimination

Within AFOAT-1 studies were underway in preparation for the potential technical talks, and to add factual material to the U.S. interagency debate on policy toward a test ban. For example, my group contributed "A Study of Seismic Methods for Identification of Explosions" in October 1957. This report concluded, among other things, that there were no known positive seismic identification criteria for explosions per se. However, we pointed out that large seismic events in aseismic regions could be viewed with suspicion, as was the case for explosions at the almost aseismic Semipalatinsk Test Site. *Rainier* and chemical explosions of about 500 tons at Promontory Point, Utah, were our primary explosive sources. Without a definitive explosion "signature," the approach we described was to identify and eliminate as many earthquakes as possible from consideration as blasts, leaving as few potentially suspicious events as possible.

The use of the term "signature" in early monitoring efforts may well have originated through involvement of U.S. Navy laboratories in the earliest days of the Long Range Detection Program. During World War II, sonar operators had demonstrated good success in identifying acoustic signatures unique to many types of ships by listening to the underwater sounds they emit. While stemming from real physical differences, of course, sonar methods were grounded in hundreds or thousands of signal samples, transmitted through a far simpler medium than the solid earth. Furthermore, unlike the seismic case, there were no natural sources of similar sonar signals to cause false alarms.

Our seismic alternative was to search for physically meaningful diagnostic characteristics of earthquakes, which abounded as signal sources. I stress *physically meaningful*: geological and geophysical conditions at potential sites for a test are so variable that there is little hope for an adequate statistical sampling. The most robust method was determination of focal depth. At the time, the method was generally accepted as valid for identifying quakes that occurred deeper than 50-70 km below the surface, and it depended on detecting seismic waves that had been reflected from the earth's surface almost directly above the focus. The interval between P and its reflection is a unique function of focal depth and distance to the detecting station, so the conclusion as to the nature of the event, when adequate data were available, was based on clear, well-understood physical principles and hence were very positive. The method was quite effective in parts of Asia, notably in the Soviet Union's far eastern and most seismically active region, but would be of little use in the United States (where most quakes are too shallow) except in Alaska and the Aleutians.

The most general of the identification methods was based on determining the direction of the first motion of P-waves. Perry Byerly laid the physical foundation for this method. His early attempts to deduce the forces at the focus of an earthquake led him to merge a Japanese theory on seismic radiation from a "force-couple" with the "elastic-rebound" theory developed to explain the great San Francisco earthquake of 1906. He predicted that sudden slippage along a fault will radiate initial compressional motions and initial rarefactional motions in alternate quadrants, separated by the fault plane and a plane perpendicular to it (see Figure 4.1). This pattern was subsequently observed experimentally around earthquakes, establishing the validity of the concept. Reduced to the simplest of terms, when a fault breaks, and the earth lurches forward in one direction, it will initially push the earth away (compressional motion) in that direction. In the opposite direction, the earth will initially be pulled inward (rarefactional motion) toward the epicenter.

Since there seems to be no way an explosion can cause initial rarefactional motion (inward, toward the source) the method should uniquely identify earthquakes. The converse is not true. If only compressional first motion is observed, this may mean only that there were no seismic stations in the directions where initial rarefactions were radiated—the event could be either an explosion or an earthquake. This would be unlikely in the case of a very large quake recorded by numerous well-distributed seismic stations; but we believed we could expect only relatively sparse station networks, and incomplete azimuthal distribution of stations, considering the size of the areas of the world where there would be no stations, only regions covered by oceans or behind the "Iron Curtain." Thus we expected the method to be successful for only a fraction of the detected events.

The third method proposed for identifying earthquakes, and the one believed by most seismologists to be the most effective, was the existence of shear waves. Our initial ideas had conceived of an underground atomic explosion as a purely compressional source—a spherical "bubble" of high-pressure vaporized rock that pushed the surrounding rock directly outward—with no side-to-side (shearing) motion. An earthquake, on the other hand, represented slippage along a fault—a predominantly shearing motion. However, atomic tests from *Trinity* to *Rainier* had demonstrated clearly to those of us who studied explosion signals that explosions do create shear waves. The reasons were far from clear. Ideas die hard, however, and we in AFOAT-1 continued to hope for a practical criterion based on shear-wave amplitudes, and thus had suggested that "unusually large" shear waves might prove to be diagnostic. There were insufficient data to quantify such a criterion, however.

Behind these purely seismological questions there lurked an even more difficult question. Even if the seismological indicators pointed strongly to an explosive origin, how could one be certain the event was nuclear? After all, the Soviets were known to have conducted underground chemical explosions of several kilotons, and the Chinese had already carried out a chemical explosion of eight kilotons. This question had been moot before the *Rainier* explosion, for all previous nuclear explosions had produced air-borne radioactivity. While the thinking was far from crisp and clear at the time, it was conjectured that some sort of visit to the site of any suspicious event, by technically trained inspectors knowledgeable of nuclear testing requirements might reveal telltale clues that a test had taken place. Even if all radioactivity were contained deep underground, the inspectors could drill for samples, or so it was thought.

Sputnik and U.S. Reactions

At about the same time, a new dimension was added to the considerations of the U.S. Government. On 4 October 1957, the USSR launched "Sputnik." The worlds' first artificial satellite. This 84-kilogram satellite achieved earth orbit and circled the earth in about 96 minutes. It was launched in support of the International Geophysical Year[10], but its regular beeps in the radio broadcast band could be heard virtually worldwide when it passed overhead, and carried an obvious non-scientific message. It was followed four weeks later by Sputnik II, which weighed a half-ton, and carried a live dog, "Laika." Who can forget the sight of that fluctuating light,[11] brighter than Venus, sailing silently across the twilight sky—its source capable of carrying a nuclear weapon? For the first time it seemed that American cities could be destroyed by a foreign power operating from another continent, with no possibility of a defense.

The U.S. attempted to place its own two-kilogram satellite into orbit in support of the IGY, but the rocket exploded on the launch pad, and the attempt failed. As was revealed many years later, President Eisenhower was aware that the U.S. had, or would soon have, the capability to place large satellites into orbit; but the capability resided in highly classified ballistic missile programs, and the President was not tempted to make counter claims. Instead, he asked for the advice of a group of highly respected scientists as to what a proper response might be. As a result, in November, President Eisenhower appointed a Special Assistant for Science and Technology, Dr. James R. Killian, Jr., President of the Massachusetts Institute of Technology and he formed a President's Science Advisory Committee (PSAC), as well.

The President also addressed the nation about the significance of the Soviet achievement, and he stressed the importance of science to national defense. He reported that he and the Secretary of Defense had agreed that, whenever possible, future missile or related programs would be put under a single manager without regard to the separate military services, and this manager would work closely with the new Special Assistant for Science and Technology. To implement this policy, in early January 1958 the President proposed the establishment of an Advanced Research Projects Agency (ARPA) within the Department of Defense. Funding and authority to perform research and development were approved by Congress and signed into law about a month later (Cole *et. al.*, 1978). The new agency, ARPA, would soon evolve into the lead agency for the U.S. government in conducting research and development in nuclear test monitoring.

The Bethe Ad Hoc Working Group

On 6 January, 1958 President Eisenhower approved a National Security Council recommendation to form a technical working group to evaluate the consequences of a nuclear test cessation. The "Ad Hoc Working Group on the Technical Feasibility of a Cessation of Nuclear Testing" was duly formed under Dr. Killian's direction, chaired by Professor Hans Bethe, a theoretical physicist from Cornell University. The newly formed working group had a number of members who would play direct roles in discussions with the USSR, to follow later that year, including: Harold Brown (University of California Radiation Laboratory and later, Secretary of the Air force and Secretary of Defense), Doyle Northrup (AFOAT-1), J. Carson Mark (Los Alamos Scientific Laboratory), and Herbert "Pete" Scoville (CIA)

Hans Bethe was widely known for his theories on the source of energy of the sun and stars, (primarily fusion of hydrogen under the intense temperature deep within the star) and for which he was later awarded the Nobel Prize. During the war he had been head of the Theoretical Division at the Los Alamos Scientific Laboratory. He was also well known to us in AFOAT-1 as the chairman of an AEC/DOD panel that performed the final evaluation of data that we had collected from foreign nuclear explosions; it was known as the "Bethe Panel."

The Bethe Working Group included no seismologist, as several writers have pointed out, although I participated in the group's meetings on February 5 and 6 in room 272 1/2 of the Executive Office Building. I remember also attending a follow-on meeting at Los Alamos, and I was the principal author of the seismic portions of an annex prepared by AFOAT-1 for the Working Group's final report.

Bethe was an excellent choice as chairman. Not only had he thought deeply about the technical and military consequences of a cessation of testing, he was also willing to listen to others whose views differed from his own. Everyone who cared to speak had their say. The main focus of debate within the working group was on military and strategic advantages each side would have in the event the U.S. and USSR should both cease testing. Strategic-military matters considered by the group included: Soviet superiority in rocket thrust and hence payload weight and the corollary need for the U.S. to develop more sophisticated weapons to achieve the same effect at lesser weight; the U.S. need to develop low yield clean weapons for possible use tactically in allied countries that might be invaded by the USSR; and deterioration of the national weapons laboratories and of stockpiled weapons. It was argued by some that these latter problems were asymmetrical: the U.S. would lose key scientists unwilling to stay with inactive laboratories, while the USSR could maintain its laboratories' key staff members simply by ordering the scientists to continue working there. And U.S. needs for more sophisticated weapons made stockpile reliability a greater challenge for us.

Verification of compliance with an agreement was also an issue, of course. Doyle Northrup presented a comprehensive overview of the capabilities and limitations of the existing AEDS, at that time based on seismic, acoustic, and EMP detectors, backed by the collection of radioactive debris. He reported that 45 Soviet tests had been detected to date. (We now know that there had actually been 51 Soviet tests, including a number in the low- and sub-kiloton ranges. From this we can infer that the AEDS had a detection threshold for atmospheric explosions near one kiloton at that time.)

There was extensive discussion of capabilities for detecting and identifying underground tests, and it became clear to all present that capabilities for monitoring underground explosions were not as advanced as those for atmospheric explosions. Potential ways to carry out clandestine tests to evade verification measures, were discussed and debated for the first time, as far as I know, by knowledgeable scientists led by Harold Brown. Among these ideas at that early stage of thinking about evasion, was simply to test underground in a remote seismic region, keeping signs of human activity to an absolute minimum. As just one among numerous other detected, but unidentified seismic events it was unlikely to be inspected, and even if it were, the inspectors would be unlikely to find the event site. There was also discussion of testing in rubble or other unconsolidated material rather than the well-consolidated volcanic tuff in which *Rainier* was fired; in this case, it was argued, the seismic signals would be reduced by absorption, perhaps by as much as ten times.

Verification discussions led to a general agreement that small tests could be hidden but large tests could not; that China, at that time closely allied with the USSR, must be included in any test ban agreement to preclude the USSR from conducting "proxy" tests there; that conducting effective on-site inspection would be very difficult; and that proving the nuclear nature of an event by drilling at the site of a deep underground test site was "practically hopeless" because of inaccuracies in seismic locations. The last point was based on difficulties that had actually been experienced in finding nuclear debris from *Rainier*, even though *Rainier*'s location was accurately known.

On 18 March 1958 a "Report on the Detection of Nuclear Tests" was sent to Bethe by Doyle Northrup for use by the Bethe Working Group in preparing its report. It contained current estimates of the detection capabilities of the Atomic Energy Detection System for tests in the USSR and China under various conditions. For estimating seismic capabilities, we assumed that a clandestine test might take place in a way that generated no acoustic or electromagnetic pulse (EMP) signals, and released no radioactivity into the atmosphere. Under these conditions, we estimated that the AEDS could detect, with high confidence (90%-100%), four kt in summer and eight kt in winter, (when there would be higher winds in the northern hemisphere, increasing the seismic noise level). Fair capabilities would exist for explosions half as large. These estimates were highly uncertain, based primarily on a single underground nuclear explosion and an assumption that seismic signals at long range could be typified by those from *Rainier*. The threshold for identification would be substantially higher, (except for tests in aseismic regions), but we had no way of quantifying it. Implicitly, a small, well-contained underground test would be recognized as such only through auxiliary information of some kind.

To lower the detection threshold to one kt, we proposed installing 43 stations inside the USSR and China. These stations were each to be equipped with an array of 20 short period sensors and three-component long period sensors. We estimated that the system would have to contend with about 2,100 earthquakes annually in the USSR and China, including border regions (remember, location errors of 50 km were considered the norm, so the effective borders were 50 km beyond land, including the highly active Kamchatka-Kurile region). These earthquake numbers, to be discussed more fully later, basically came from publications of Gutenberg and Richter. It was estimated that about 300 events annually would remain unidentified seismically.

To supplement the AEDS in monitoring the remote regions of the Earth, about 10 similarly equipped seismic stations were proposed. The combined system, however, would have a detection threshold of about 20 kt in remote regions, and about 400 seismic events would be recorded, but unidentified annually.

On 27 March 1958 the Bethe Ad Hoc Working Group Report was submitted to the President. An AFTAC report on detection of nuclear tests had been adopted by the Ad Hoc Group, and appended to their report. A range of opinions was reported on the Group's major issue; i.e., whether a cessation of testing would be to the net military advantage of the U.S. However, they did agree that the U.S. should not stop testing until the completion of the test program then under preparation, Operation Hard-tack. On the question of verification, they found that a system could be designed to detect nuclear explosions except small (1-10 kt) underground tests. For this purpose, they proposed a system with 70 monitoring stations inside the USSR and China. Shortly thereafter the report was reviewed by the President's Science Advisor and by the PSAC, with the result that Dr. Killian reported to the President that a suspension of testing at the end of 1958 would be in the country's best interest.

The Senate Foreign Relations Committee's Hearings

During the late winter and early spring of 1958 the Subcommittee on Dis-armament of the Senate Foreign Relations Committee conducted hearings on the suspension of nuclear tests. Much of the testimony was classified, but much of the substance leaked to the press, and a "sanitized" commit-tee report was soon published, reflecting conflicting views among scien-tists as to the wisdom and the verifiability of a test suspension. None of the witnesses had direct experience at test detection and identification. Dr. Bethe was well informed from his Working Group experience, and reflect-ed the test detection conclusions of his recent report. Other witnesses were less versed in the subject, but nevertheless provided opinions, including such misinformation as that a seismic system could "pinpoint" the loca-tion of a suspicious event "within about a mile."

The leading scientific voices were those of Bethe, who advocated a suspension and argued that verification was possible, although difficult, and of Edward Teller, who opposed suspension. Teller argued the need for "clean" weapons (low fallout) and the importance of low yield tests. He went on to argue the difficulty of identifying small tests, and the pos-sibilities of evasive testing methods, which were essentially unexplored at the time. The major emphasis of the hearings was on weapons and the significance of suspending tests, but the importance of verification was made clear to the Committee, as was the wide range of opinion on how well it could be done. Committee members expressed their confusion en-gendered by the contradictory testimony they had heard.

The Subcommittee on Disarmament and its Questionnaire

Senator Hubert Humphrey, the Chairman of the Subcommittee, decided to try to clarify matters. On April 10, 1958, he sent a letter and questionnaire "seeking the opinion of a number of seismologists" on "our technical capability to detect and identify underground explosions." The questionnaire contained 19 multi-part questions broadly covering the subject. My reaction at the time was, and still is, that an opinion poll is the wrong way to address scientific questions -- but I responded anyway, as did 30 others. The responses were subsequently published as Staff Study No. 10 of the Subcommittee on Disarmament. For security reasons, I was identified in that study by the title "Geophysicist, former Chief Seismologist, Geotechnical Corp., Dallas, Texas."

Several seismologists declined to answer some or all questions, citing their lack of critical information, or their need to conduct extensive research before giving a responsible answer. In my cover letter I cautioned that four areas appeared to be particularly uncertain to me. These were:

1) *Detectability of blasts larger than* Rainier *(1.7 kilotons)*—While the amplitudes of earth motion produced by *Rainier* are fairly well known, there does not appear to be a well-established theory by which to scale up from 1.7 kilotons to 10 kilotons, or larger.

2) *Dependence of seismic wave amplitudes on material surrounding an explosion*—There are no established relations to use in estimating the size of seismic waves from a *Rainier*-type bomb exploded in granite, salt, or material other than tuff.

3) *Methods for discriminating between earthquakes and blasts*—As a result of an intensive study of the seismic waves from *Rainier*, I have concluded that the best means for differentiating between shallow earthquakes and underground blasts is through an analysis of the first motion in the P-waves. . . A seismic inspection system could thus identify many, if not most, earthquakes but there would remain a number which are totally unidentified . . .

4) *Statistics on small shocks in the USSR and China*—For my own estimate, I have used Gutenberg and Richter's statistics[12] for "Class A" shallow shocks and assumed that smaller shocks in the USSR and China are in the same proportion as in the world as a whole. . . All estimates are subject to wide variations from year to year, and my belief is that a realistic inspection system must be prepared for a time of high seismic activity rather than "average" activity if the desire is to prevent weapons development.

I concluded: "The only way I can suggest to settle the first three of these problems is to conduct a series of underground nuclear explosions.

After such tests, seismologists will be much more unanimous in their views than will appear in the set of divergent answers to this questionnaire which I feel confident will result from the individual seismologist's evaluation of the small amount of positive information available at the present time."

However, Senator Humphrey, a strong advocate of a test ban, found considerable comfort in the collective responses. In a preface to Staff Study No. 10, he concluded "The weight of evidence produced by this study . . . (full title) . . . indicates that an inspection system for a nuclear test moratorium could provide a high degree of assurance that no country would risk violating the agreement by trying to conduct tests in secret."

Comments on Findings in Staff Study No. 10

Rather than providing scientific support for Senator Humphrey's sweeping conclusion, I think the study's true value is that it provides a benchmark on what was known (and not known) at the time—as we approached technical discussions with the USSR.

The Staff Study, dated June 23, 1958, contained a Summary Analysis, reporting answers from the respondents, as outlined in the following:

- Seismologists with experience with deep underwater explosions reported them to be easier to detect and identify than underground explosions.
- There was wide disagreement among the respondents on the maximum detection distances for underground explosions, as shown in Table 4.1.
- The distances at which the explosion could not only be detected, but located, showed similar differences in estimates among the respondents, but the distances were somewhat smaller. Location accuracy would vary with the number of detecting stations, but errors as small as 5-10 miles were frequently mentioned.
- Estimates of the size of an explosion which would generate a magnitude 6 event ranged from 44 kt to 2,000 kt. I replied that the relationship between magnitude and yield had not yet been established. (However, assuming 1/200th of the energy went into seismic waves, as was estimated for *Rainier*, the required yield would be about 1 Mt).
- Estimates of the number of earthquakes per year in the USSR and China between magnitude 4 1/4 and 6.0 ranged from 50 to 2,000.

On the questions concerning the critical matter of distinguishing between explosions and earthquakes the study reports a wide range of opinion. Several respondents were very positive, reporting that explosions can be unambiguously identified—one asserting that detection of signals at only a single station would be sufficient. Others flatly rejected the possibility of unambiguous identification of explosions. However, most agreed that there were circumstances under which there was a good probability of identification. Several mentioned determination of focal depth as a criterion. The direction of first motion in P waves was also frequently mentioned, with rarefactions cited as an identification feature for earthquakes. The replies reflected almost complete agreement that it was not possible from seismographic data alone, to distinguish between chemical and nuclear explosions (but the size might be a clue).

Other "distinguishing characteristics" singled out by Senator Humphrey's staff study included:

1) A blast produces a greater percentage of surface (L) waves compared to body waves (P and S) than an earthquake.
2) The S waves from an explosion have smaller amplitudes relative to the P waves as compared to the amplitudes of the cor responding waves from earthquakes,
3) Explosions produce compressional seismic waves of very high frequency compared to earthquakes.
4) The times of arrival of the signals at near and far stations reveal if the source was at or near the surface, and earthquakes usually are deeper than explosions.
5) An explosion is a point source whereas an earthquake has an extended source and the overall duration of the disturbance is less for dynamite blasts than earthquakes.

Well, considering the range of *opinions* expressed, the senator's staff might have had difficulties sorting through the various views expressed. One characteristic (1) we now know the opposite is true; i.e., the ratio of surface waves to body waves is larger for *earthquakes*. Seismograms from *Rainier* had shown characteristic (2) to be untrue, at least as a generalization. Characteristic (5) is a truism, describing differences in the nature of sources, but useless for identification purposes. Other features cited, while not distinguishing features by themselves, have at least elements of validity, some of which are still being worked on today.

Other points of interest included:
• The estimated number of stations required to monitor the USSR for explosions down to 1 kt ranged from 40 to 100.
• Testing methods that would make detection and identification difficult included testing in seismically active regions, or near active mines, and timing the test to occur shortly after a large earthquake or mining

blast; testing in an absorbent material; and testing in a drilled hole under a shallow sea or lake to foil on-site inspection.

For the most part, the range of opinion in the replies were testimony to the existence of large areas of uncertainty, and that many qualifications applied to the answers given. Many of the questions were the same ones some of us would face from Soviet scientists if the technical meetings proposed by the President were to materialize. The wide range of answers given and the uncertainties they expressed might well have prompted the President to conclude that additional work might be prudent before technical meetings. The meetings had been agreed to, however, before the study was released.

A Proposed Seismic Monitoring System

The first public description of a seismic monitoring system for the USSR by a knowledgeable seismologist seems to have been given by Professor Frank Press. Press recently had been appointed Director of the Seismological Laboratory at Caltech when Beno Gutenberg stepped down from the position. He was a young, rising star in his profession who had studied geophysics under Professor Maurice Ewing at Columbia University's Lamont Geological Observatory. Press had responded to Senator Humphrey's questionnaire, and obviously he had given careful thought to his subject. As reported in the *Washington Post* and *Times Herald* on May 1, 1958, he had proposed at the annual meeting of the National Academy of Sciences that 100 seismic stations in Russia could monitor nuclear testing, and at the same time provide fundamental information for science. This proposal echoed his response to Senator Humphrey, in which he suggested equipping the stations with sensitive short period seismometer arrays, as well as long-period seismographs. Explosions down to about 2 kt could be successfully monitored, according to this report. Press later became President Jimmy Carter's science advisor as head of the Office of Scientific and Technology Policy on the White House staff, and ultimately he became President of the National Academy of Sciences.

Agreement on Technical Talks

Agreement to hold technical talks on a nuclear test ban had finally come about through an exchange of letters in April and May of 1958 between President Eisenhower and the latest Soviet Premier, Nikita S. Khrushchev. The exchange began less than two weeks after completion of a major series of Soviet nuclear tests, and during the early stages of a highly publicized U.S. test series. Khrushchev proposed that the U.S. announce a moratorium on tests to match a moratorium declared by the Supreme Soviet (after completing their test series). Eisenhower noted the "peculiar" timing of this proposal in his response, but he also reiterated a recommendation made to former Premier Bulganin in January on technical discussions between the two sides. Khrushchev responded with familiar words that special technical monitoring means were not needed; but finally, to the surprise of many, he agreed on 9 May 1958 to appoint experts to begin technical discussions. Further correspondence in June established that a Conference of Experts would meet in Geneva on 1 July to study "methods for detecting possible violations of an agreement on the cessation of nuclear tests."

Dr. James B. Fisk, then Vice President of the Bell Telephone Laboratories and a member of PSAC was announced as chairman of the U.S. delegation on June 20. Other delegates were Professor Robert F. Bacher, Chairman of the Division of Physics, Mathematics and Astronomy at Caltech, and Dr. Ernest O. Lawrence, director of the University of California's Radiation Laboratory. Lawrence was renowned for his invention of the atom-smashing cyclotron and for his use of it to explore the structure of the atom. For this he had received the Nobel Prize in physics in 1939. Bacher had participated in the development of the A-bomb at Los Alamos, and had served as a Commissioner of the Atomic Energy Commission. Fisk had served as director of research for that Commission, and earlier, had been a member of the Boner Panel. Their selection as a group had been designed to represent a wide range of viewpoints on test ban issues to insure objectivity.

The agreement between President Eisenhower and Premier Khrushchev also called for participation of other countries allied to the U.S. and USSR. The eastern delegations were to include experts from Poland, Czechoslovakia and Rumania, while the western delegations would include experts from the United Kingdom, Canada and France. The experts from these eight countries would be tasked with laying the technical foundations for nuclear test ban treaty negotiations.

FIVE

The Conference of Experts
The Seismic Method

Much has been written about the Conference of Experts, in part because it broke international political precedents in three ways: first and foremost, it was the first time scientists had been given a specific, independent negotiating task on behalf of their governments. True, international meetings had long been a tradition among scientists, but participants represented only themselves, or perhaps scientific institutions, but not their governments, at least not among Western nations. Second, parity between East and West was accepted in the number and political orientation of the countries represented. This contrasted with procedures of the United Nations that were based on a "one country-one vote" principle, not on equal representation between Eastern and Western political blocs. And third, it took place outside of the U.N., reportedly causing some concern to Secretary General Hammarskjold. However, he offered U.N. facilities and services in Geneva for the meeting, and these were accepted.

It seems to me that another, less noticed, precedent was also set: the establishment of scientific verification methods and on-site inspection as key elements of future arms control agreements. It was far from clear as we began the meeting that agreement could be reached with the Soviet Union on either element. But as we will see, both were agreed upon, and a scientifically based system for monitoring nuclear testing was described in the final report of the Conference, as was the need for on-site inspection.

The Western Delegations

In a formal sense, the U.S. delegation to the "Conference of Experts to Study the Methods of Detecting Violations of a Possible Agreement on the Suspension of Nuclear Tests" were the three scientists previously named, headed by Dr. James Fisk, who also headed the four nation

Western delegation. Supporting the U.S. delegation, there were a number of advisors, selected in part on the basis of U.S. Governmental Agency representation, but most of them on the basis of technical expertise. Advisors who would play a role in seismological discussions included Hans Bethe and Doyle Northrup, whose roles in U.S. preparations leading up to the conference have been discussed previously; Dr. Harold Brown (University of California Radiation Laboratory); Dr. Richard Latter (Rand Corporation); and, of course, the seismologists. These were: Professor Perry Byerly (University of California), Dr. Norman Haskell (Air Force Cambridge Research Center), Professor Jack Oliver (Columbia University), Professor Frank Press (Caltech) and myself,[13] identified only as with the "U.S. Air Force"—the AFOAT-1 affiliation being omitted for security reasons.

Delegates from other Western nations were: Sir John Cockcroft and Sir William Penney from the Atomic Weapons Research Establishment of the United Kingdom; Professor Yves Rocard of the Laboratory of Physics, University of Paris, France; and Dr. Ormond Solandt of the Defense Research Board of Canada. Sir John Cockcroft was a legendary figure to me, famed for his Nobel prizewinning work in transmuting atomic nuclei by bombarding them with protons. At the time, he was Director of the Atomic Weapons Research Establishment at Harwell. Sir William Penney had participated in the development of the atomic bomb along with U.S. scientists at Los Alamos, New Mexico in 1944-45. At the time of the Conference, he was Director of the Atomic Weapons Research Establishment at Aldermaston.

Sir William Penney and Professor Yves Rocard were experienced in various aspects of nuclear test detection (among other more primary areas of expertise) and were the principal scientists of their countries who were actively cooperating with AFOAT-1 to monitor testing by the Soviet Union. Cooperation between the U.S. and the U.K. and Canada dated from the beginning of work on the bomb, and was extended to test detection in the early days of the Long Range Detection project. It was a different matter with France. Faced with the determined efforts of President Charles de Gaulle and others to create a French nuclear force that was independent of NATO, official U.S. policy was to avoid any form of cooperation on nuclear matters. Consequently, cooperation on test detection proceeded only on a very limited and discrete basis until some years after the first French nuclear test in 1960. Test detection stations in France had been established, independently from AFOAT-1 by Yves Rocard. (Michel Rocard, a future Prime Minister of France, was his son.)

Guidance

Instructions to the U.S. delegation were quite broad and general, and seem to have been verbal—I recall no instance of referral to a written "terms of reference" or similar document. However given, the elements of the guidance were: to seek common understanding of the techniques for detecting nuclear tests; to outline possible systems for monitoring a possible treaty; to analyze and describe the capabilities and limitations of these systems; and to use our best scientific judgment. No goals were set on such fundamental matters as what size explosion the systems should be able to monitor and no advice was given on how the results would be used by the U.S. Government, once a report was produced.

For most of us the essence of this guidance was received during our initial meeting as a group with Jim Fisk in the U.S. Consulate in Geneva. I had been briefed on the meetings a week or two earlier in Washington, D.C., by some representative of the State Department, but most of what I remember about the briefing had to do with security (don't talk in hotel rooms, etc.), and that I would need a tuxedo for inevitable soirees with the Eastern delegations. Utter nonsense, the latter! The Russians and other eastern European scientists had no such elegant attire; possibly it was not even available to them in Eastern countries still not fully recovered from a devastating war a dozen years ago. My tux was never worn in Geneva —but it still fits!

The Conference Convenes

The Soviets kept the U.S. and its associated countries guessing whether they would actually attend almost up to the moment of convening the meeting. For several months they had argued for a prior commitment by the U.S. on a cessation of testing. The U.S. insisted that the talks were to take place without preconditions. The Soviets maintained the stance that if the U.S. did not agree "that results of the experts meeting should assure the cessation of tests, it is useless to send experts. In this situation, the USSR cannot send experts, and will not be an accomplice to deception of the people." Secretary of State John Foster Dulles sent us to Geneva anyway, with the matter still unsettled to the best of our knowledge.

The Conference of Experts began on Tuesday, 1 July 1958 in a grand conference room of the *Palais des Nations* (see Figures 5.1 and 5.2), originally built to house the League of Nations. I remember entering the Palais feeling excitement at this unique event—and some apprehension caused by the unusual responsibility we had shouldered—heightened by uncertainty about whether the Eastern scientists would show up, or if they did, who they might be.[14] But they did show up, and after brief, ceremonial greetings among the delegates at the table and a few minutes for photographers and members of the press, the conference was underway. As had been agreed among the parties, the conference took place behind closed doors, facilitated by an expert staff of simultaneous translators and stenographers. Verbatim records were printed daily in English, Russian and French, and these greatly aided those of us who were used to absorbing most information through our eyes, and perhaps not as well through our ears. Brief, bare-boned communiqués on the work of the Conference were released to the press at the end of each meeting.

The Eastern Delegations

Later, in a "post-mortem session" at the Consulate, we were able to compare notes on what we knew about several of the Soviet delegates whom we recognized. Dr. Yevgeni K. Fedorov, the leader of the Eastern delegations and a member of the Communist Party, was Director of the Soviet Academy of Sciences' Institute for Applied Geophysics, and was prominent in the ongoing International Geophysical Year (July 1957–December 1958). Dr. N.N. Semenov, a Nobel Prize winner, and Dr. I.Y Tamm were well known and highly respected physicists. Dr. Ivan P. Pasechnik was known to have published on seismic prospecting, but an entirely different field of more recent work was revealed to us when the seismic discussions began.

Two differences in the composition of delegations of the Soviet Union and the United States soon became apparent. First, the USSR had included a senior diplomat, S.K. Tsarapkin, who had served in many important posts representing the USSR. And second, the USSR had included nuclear detection experts as formal members of their delegation. These experts were Drs. M.A. Sadovsky, O.I. Leipunski, I.P. Pasechnik, and K.E. Gubkin, as we were to discover. Aside from Pasechnik, the others were virtually unknown to us initially.

The only non-Russian that we knew was Dr. A. Zatopek of Czechoslovakia, some of whose seismological publications I had read as a graduate student. Except for Soviet scientists, he was the only member of the Eastern delegations who participated in the Conference in any substantial technical way.

Later, when seismic discussions began, the Soviet delegation was supplemented by three other advisors: Drs. L.M. Brekhovskikh, V.I. Keilis-Borok, and Yu. V. Riznichenko, all senior geophysicists whose important publications were well known to us. Riznichenko had been elected a Vice President of the International Association of Seismology and Physics of the Earth's Interior at the Toronto meeting previously mentioned, and in all probability, had attended the impromptu meeting on planning for

Rainier. Most of the seismologists who attended the Conference are shown in Figure 5.3.

In what follows I will make no distinction between "delegates" and "advisors"; we functioned as a unified U.S. delegation, as I believe did the delegations of other countries.

The presence of Mr. Tsarapkin on the USSR delegation, as well as senior diplomats on the delegations of Poland and Czechoslovakia, signaled an important difference between our understanding of the purpose of the meeting, and that of the Eastern group. Our approach, to attempt to keep the discussions as objective and scientific as possible, was soon to be tested.

It has been suggested (e.g. Gilpin, 1962) that the lack of a senior diplomat on the Western delegation was a disadvantage. I don't think so, at least not during the Conference itself. Among our several diplomatic advisors was Ronald I. Spiers of the Disarmament and Atomic Energy Affairs Office of the U.S. Department of State. While junior to Tsarapkin, he was well versed in U.S. nuclear policy and disarmament affairs as well as in the procedures and protocols of international interactions. He was also highly perceptive of diplomatic/political nuances that others of our group might miss or misunderstand. Our objectives were to provide scientific advice to our government, where political decisions would be made, and Spiers was an excellent choice to support the technical people, it seems to me.

It is certainly true that our "scientific advice" was set in a political context, where Tsarapkin could, and did, provide instant advice to his delegation. But on important questions we also had advice available from senior U.S. Government officials, who followed the Conference closely through our daily-cabled summaries. After the Experts' Conference, Tsarapkin became the USSR's ambassador to test ban treaty negotiations, later in 1958. There, his knowledge of the means and complexities of test ban monitoring, as revealed to him by the Experts, and especially of the contents of the Experts' Final Report, proved to be a significant advantage.

The Agenda

The first business before the Conference was an attempt to define an agenda for the Experts. Fedorov promptly raised an essentially political issue by proposing a statement of objectives that amounted to a prior commitment to a test ban. This was precisely the issue previously rejected by the U.S. and over which the USSR had almost balked at attending the Conference—the venue for discussion had merely shifted from the diplomats to the scientists, as Fedorov saw it. Fisk countered that the Conference should confine its work to defining methods of detection, analyzing their capabilities and limitations, and let Governments decide how to use the information. Fedorov was persistent, and I recall seemingly endless repetitions of words like "But what is our objective, Dr. Fisk? Surely our time is wasted if our work is not intended to secure a test ban." Fisk was equally persistent, and tensions mounted in the conference room. It appeared likely that there would be no discussions of technical matters and the conference would collapse. Tensions were so high, and the outlook for the conference so bleak, that E.O. Lawrence prepared a letter to his Nobel counterpart on the Soviet delegation appealing for his support in continuing the talks. (It seems to remain a mystery as to whether the letter was actually delivered, but there is no doubt that it was prepared, and that Lawrence felt strongly about its contents.)

An agenda was adopted on Friday, 4 July 1958 after a one-day recess, during which Moscow apparently changed its guidance to the Soviet delegation. The agenda contained four tasks for the Conference:

1) An exchange of opinions on various methods for detecting atomic tests.
2) Determination of a list of basic methods of systematic observations for the phenomena indicative of an explosion.
3) A system combining the various methods for controlling the observance of an agreement on the cessation
4) Drawing up a report to governments providing conclusions and suggestions on a system for controlling the observance of an agreement on the cessation.[15]

Technical Talks Begin

As the Conference turned to its technical work, it began to consider each of the known techniques for detecting atomic explosions. These were, in order of their consideration: acoustic, including hydroacoustic methods, the collection of radioactive debris, the seismic method, and the detection of radio signals (which we in the West called "EMP," for electromagnetic pulse).

A pattern to the meetings soon evolved. There would be technical presentations by each side, usually followed by questions. Typically the U.S. or U.K. led for the Western side, with presentations describing actual experience with a given method and backed by actual data. AFOAT-1 scientists usually provided the substance of our primary descriptions of the capabilities and limitations of each method. Soviet presentations for the Eastern side tended to be more general, and the data they included tended to be less relevant. I soon speculated that we were mostly dealing with members of Soviet academia, who well understood the basic physics of each technique, and probably had made some experimental measurements of signals or debris from bombs, but had not faced the problem of applying the technique operationally to long range detection of unannounced foreign tests. This view, which I'm sure others shared, turned out to be correct in most respects; although it was only after several decades had passed that we were given a reasonably good view of the Soviet nuclear test monitoring activities and the roles of our Soviet contemporaries (see Appendix B).

Following the discussion on a given method, one side or the other would table a "Conclusion on the Applicability" of the method. The Conclusion described the underlying phenomenology, detection ranges and associated propagational characteristics, the sources and frequency of occurrence of false alarms, and other general considerations. A second part of the pattern developed by the Conference emerged at this stage: the Western scientists proposed specific numbers when possible, and ranges of numbers to indicate variability or uncertainty, while the USSR scientists proposed more general statements, rather than numbers, and focused on descriptions of capabilities that were more qualitative in nature.

A pattern to our daily lives also evolved, centered on the conference. Following each meeting, we assembled for a "post-mortem" in the conference room of the U.S. Consulate, with representatives of the other Western delegations usually in attendance. Sharing ideas on what we had heard, or thought we had heard, from the other side was always first-order business —translations by translators unused to scientific terms were sometimes inexact, and nuances could be lost. Writing assignments were made or amended as the delegation decided on our next presentation at the meetings. We then adjourned for dinner.

Excellent, affordable (then) restaurants abounded in Geneva, and various groups spread out around the city to sample them. Jack Oliver, Norman Haskell and I were frequent dinner companions, often joined by others of our delegation. We typically paused first at some sidewalk cafe for liquid refreshment, savoring the balmy evening air and observing the passing parade of walkers, while considering where to eat. I don't recall a single mediocre meal. Afterward, back we went to the Consulate to write or review papers drafted by others—often until midnight, or later.

The morning routine was to write and review, followed by a mid-morning delegation meeting, at which each of us could comment on the papers to be presented to the conference later that day. After lunch—in my case, usually a sandwich brought in from a small local cafe—we continued writing or studying for some future paper until it was time for the meeting.

This routine was punctuated by several receptions in the early evening sponsored by one or more of the Eastern or Western delegations. I remember particularly a reception to celebrate the Polish national holiday in late July. At these affairs we were able to associate with our foreign colleagues in a relaxed way, quite different from the frequently tense atmosphere of the formal meetings. I came to appreciate the Russian sense of humor — quite similar to American humor — and mere professional respect for our opponents began to evolve into acceptance of them as colleagues, and perhaps the beginnings of friendship.

Acoustic methods were the first to be discussed in the formal meetings, followed by methods for collecting radioactive debris. The latter method had contentious problems associated with it, involving the need to use aircraft for collecting radioactivity borne high into the atmosphere by the rising fireball of an atomic blast. The Soviets contended that aircraft were unnecessary; the West strongly supported their use. After agreeing to the use of aircraft, the delegation faced equally contentious political questions of allowable flight routes through one another's territory, and the national composition of flight crews. The Soviet reluctance to allow any foreign intrusion into their territory was clearly evident in the discussions. It was not until 14 July that the Conference turned to the seismic method.

Seismic Methods

I made the initial presentation on the seismic method as a result of a flip of a coin. Jim Fisk was chairman that day (the two sides alternated) and he had proposed to begin seismic discussions with my talk; Fedorov complained that the previous session had been opened by a Western expert and proposed a presentation by Dr. I. P. Pasechnik instead. Fisk reached into his pocket for a Swiss coin, remarking that there was a classic way to settle matters of that sort, and asked Fedorov to call heads or tails. "Heads" responded Fedorov, "It's tails" said Fisk, so I would start and we would "look forward with pleasure to [my] being followed by Dr. Pasechnik."

My talk was essentially a tutorial that broadly covered the seismic method. It mentioned briefly the relative coupling of explosive energy into seismic waves from underwater, underground and atmospheric blasts. It outlined what we had learned about P-wave amplitudes and their dependence on yield and distance, much of it from large explosions in our Pacific proving grounds. It described our understanding of seismic noise and its influence on detection and the design of seismographs. It mentioned the use of arrays to improve the detectability of seismic waves. It went on to report the absence of known positive seismic identifiers of explosions, but I outlined our approach of identifying as many earthquakes as possible by determining the direction of first motion and the depth of focus. I gave no estimate of the expected numbers of unidentified seismic events. Indeed, this was not possible at this stage, since it would depend in a sensitive way on the characteristics of our yet-to-be defined monitoring system.

Dr. I.P. Pasechnik then opened for the Eastern side with a lengthy talk —mostly tutorial—illustrated with a number of tables and figures, and he showed numerous seismograms. The seismograms, many from broadband recordings of megaton size surface explosions, were used mainly to support his claim that explosions were easily recognizable.

Pasechnik then took issue with Dr. Fisk's remarks of 2 July in which Fisk had pointed out the large numbers of earthquakes and the difficulty of seismic discrimination. Undeterred by the absence of knowledge of capabilities of the still undefined control system, Pasechnik voiced the opinion that the "overwhelming majority" of earthquakes each year would be excluded from consideration as possible explosions by seismic means, leaving "5 or 10, I say, which might raise some doubt."

Stretching the maximum limits of performance, or selecting the most favorable of conditions, had already become a well recognized hallmark of Soviet assertions and proposed conclusions during the Conference. As later reported by R.W. Snelling (1967), a member of the U.K. Mission to the Nuclear Tests Delegation:

> Their speakers were at pains to emphasize the efficiency of the seismic method, and hence to minimize the number of control posts that would be necessary on Soviet territory, and to minimize the need for on-site inspection. To achieve those ends, they exaggerated the probability of the detection of first motion, and gave the seismic-amplitude energy-yield ratio as high a value as possible.

Both talks were followed by numerous questions, and it was clear that significant differences existed between our two sides. It was also clear that the discussion would need to continue for a number of days.

Informal Meetings

It had become apparent that arguing each point at issue on each technique, and attempting to draft agreed conclusions in the fully assembled formal meetings, was a slow and inefficient process. This recognition led to an agreement to meet in informal sessions as well. Under this arrangement, experts in one particular technique would meet (usually in the mornings), go into as much detail as needed to explain and resolve differences, draft conclusions, and when agreed, present these conclusions to the formal meetings (usually in the afternoon). Doyle Northrup and Mikhail

Sadovsky, (by then the latter had emerged as the senior bomb detection expert on the Soviet side), would chair the informal meetings.

The first informal seismic discussions took place in the Palais on 17 July 1958. Unlike the formal meetings, there would be no simultaneous translation service by the U.N., nor would there be stenographers to record the discussions verbatim. Fortunately, the U.S. group (Northrup, Oliver, F. Press, and Romney at the first meeting) were joined by Mr. Peter Kelly and Dr. Robert Press of the U.K. Kelly was an experienced scientific Russian/English translator; R. Press had served as the atomic intelligence representative at the British embassy in Washington, D.C., where his duties included liaison with AFOAT-1.

Both Press and Kelly were members of the "Technical Research Unit," the U.K. Ministry of Defense unit responsible for analyzing intelligence on the Soviet nuclear weapons program. As its head, Press was widely knowledgeable of nuclear test detection methods. As it turned out he was also a superb note-taker. In places, his notes are almost verbatim recordings of the informal meetings, no doubt helped by the fact that everything had to be said twice (in both Russian and English). The Eastern group at the first meeting (Sadovsky, Pasechnik, Gubkin and Zatopek) were later joined by Dr. L.M. Brekhovskikh, a well-known hydroacoustics expert and theoretician, and Drs. Sokolov and Jara (both unknown to me then and now), supported by a translator.

I have the Press notes on five lengthy seismic meetings and two shorter informal meetings—continuations of meetings interrupted by formal meetings—between July 17 and July 23. Each of the Western seismologists took part when they were not preoccupied by other duties required by the formal meetings. The notes show that Sir Edward Bullard, former director of the U.K.'s National Physical Laboratory, attended several informal meetings; as did seismologists Dr. Patrick Willmore representing Canada and Dr. J.P. Rothe, Secretary General of the International Association of Seismology and Physics of the Earth's Interior, representing France. A notable exception was Professor Perry Byerly, who had returned to the United States, concerned that recent "police actions" by U.S. troops in Lebanon (to avert a threatened civil war) might trigger more general hostilities in Europe.[16] The informal meetings were conducted in a cooperative and courteous manner, and were centered on points of scientific differences— a welcome relief from the formal meetings where Fedorov always seemed to keep contentious issues before us.

Drafting Conclusions

One by one the seismological issues were aired, clarified and sometimes resolved. At least the informal sessions were conducted more nearly like scientific meetings than were the formal meetings, and we were comfortable in this environment. We continued with our work of crafting agreed conclusions.

And "crafting" is a good descriptive term for what we did. I had earlier described seismic array stations that could detect first motion from 1 kt at 1,000 km at "exceptionally quiet" stations. The Russians objected to the use of these words in our agreed conclusions and quoting from Bob Press' notes:

> Mr. Northrup said perhaps 'exceptionally quiet' means something different in the Russian language. Dr. Sadovsky commented that in Russian the word 'exceptionally' means 'almost impossible'. Mr. Northrup asked if 'extremely' would meet the point. Drs. Pasechnik and Brekhovskikh immediately said that would be worse. Dr. Sadovsky asked if we could say 'considerably more quiet than average'. Mr. Northrup agreed. Para. 2(a) was then amended...

"Crafting" the conclusions not only resulted in such awkwardness of expression, it also resulted in the loss of technical content in some cases. The Western scientists had proposed quantitative definitions of certain terms, like noise levels at "quiet" or "average" seismic sites, and we proposed ranges of yields, or distances, to indicate uncertainty about these parameters, but this was not accepted by the Eastern group. Thus our statement that "it is possible to register underground explosions of one kiloton at a distance of the order of 1,000 kilometers. However, the determination of the direction of first motion may not be possible beyond 500 kilometers for average conditions, or beyond 700 kilometers in favorable conditions..." was amended in the negotiations so that the qualifying sentence read, " ...first motion may not be possible in all cases." We attributed the reluctance to include such numbers chiefly to directions from Fedorov, but it was also argued that these numbers were not well known. True enough, but the paucity of such numbers left many statements open to later political interpretation—possibly Fedorov's intent.

Rather than attempting to describe the sometimes heated back and forth of the meetings and the conference, in the following sections I will outline what seemed to be the more important issues, and the differing views.

Note to readers: What follows in this chapter is intended primarily for those interested in the science. Others may skip to the next chapter, which gives conclusions of the Conference, without losing the main thread of this book.

Dependence of Amplitude on Distance

My opening talk at the conference had introduced our understanding of the dependence of P-wave amplitudes on distance; that is, amplitudes of the first arriving P-wave had been observed to decrease about inversely as the third power of the distance from a few hundred km to more than 1,000 km.[17] Dr. L.M. Brekhovskikh challenged this relationship, citing classical theory that predicted P-waves at distances of several hundred km should decrease inversely as the second power of the distance. Recalling Professor Byerly's principle on the relative importance of data and theory, I responded, "Yes, I am aware of the theory. I can only say that these are observations . . . and I suggest that if the theory took into account a velocity gradient at the base of the crust . . . this could modify the theory enough to account for what we observe."

I should comment that there is no theoretical basis for an R^{-3} decrease in amplitude of P-waves with distance (R). It is strictly empirical, but fairly accurate for NTS explosions and its use simplifies many calculations. The classical theory for "head waves," which we believed Pn to be, had been published by Cagniard (1939) and others; the theory predicted a more complicated dependence on distance (but roughly a R^{-2} decrease). If further corrected for absorption by a term of form e^{-kR}, where k is empirically determined by the data, one can also obtain a good fit to the observations (Pasechnik, et. al., 1960, Wright, et al., 1962).

I also drew attention to the low amplitudes in a "shadow zone" beginning somewhat before 1,000 km and continuing to about 1,500 km from the sources, with higher amplitudes observed at both greater and lesser distances. To illustrate this, I showed a seismogram and presented the amplitude vs distance curve for a one Mt surface explosion as we had inferred it from observations of H-bombs in the Pacific (Figures 5.4, 5.5). I pointed out that amplitudes beyond about 3,000 km were uncertain, since observations could only be made at distances where islands accessible to our experimental stations existed. Furthermore, we could not be certain in all cases that our observed amplitude changes reflected systematic distance effects, rather than path- or station-related amplitude anomalies.

Pasechnik proposed a formula for estimating the P-wave amplitude of underground explosions as a function of both yield and distance. It was apparently based on the assumption that the amplitude of the P-waves decreases as the first-power of the distance and is directly proportional to a "constant" times the cube root of the yield.[18] He had estimated the constant, C, from a few amplitudes observed from underground chemical explosions of 50 tons, 1,000 tons and 3,100 tons in the USSR, and an assumed amplitude from the one underground nuclear explosion, *Rainier*. We recognized the larger two chemical explosions as the Arys' explosion of 19 December 1957, near Tashkent in Central Asia, and the Pokrovsk-Ural'skiy explosion of 28 March 1958, just east of the Ural mountains. The latter had been detected at nine conventional stations outside of the USSR, and both had been well detected by AEDS stations.

Pasechnik acknowledged that his method was crude, but felt that it could be used for our purposes. He used this relationship to calculate amplitudes for 10 kt and 100 kt at distances of 1,000, 2,000, 4,000 and 10,000 km. And while he didn't mention it, the relationship predicted that *Rainier* should have generated amplitudes larger than about 1/2 micron at all distances out to 4,000 km, a signal well above the detection threshold at virtually every seismic station in the U.S. and Canada! A far cry, indeed, from our actual observations that showed that numerous stations beyond 400-500 km had not detected P-waves from *Rainier*. Pasechnik amended his statement the following day, stating that his relationship applied only between about 1,000-2,500 km. But even then, he said, "I don't want to press it," and it was dropped quietly by the Conference.

The Regional Shadow Zone

The concept of a shadow zone near 1000 km, which seemed to fit our observational data, became a contentious issue between the two delegations. The Soviets reported that amplitudes between 900-1100 km were almost 100 times larger than we observed from *Rainier*, apparently based on the previously mentioned chemical explosion in the Urals (the experimental basis for this statement was not well-described during the conference, much less supported with verifiable seismic data). In an informal meeting, Brekhovskikh gave a scholarly lecture on seismic ray paths through the earth's upper mantle based on an assumed velocity versus depth model that did not include Gutenberg's low velocity channel, the suspected cause of the shadow. Instead, Brekovskikh's hypothetical earth structure had regions where the velocity increased suddenly with depth, which would create zones in which the P-waves were focused at the earth's surface, and this might explain the large amplitudes. Frank Press responded that we must base our conclusions on experimental data, and the data available to us defined an amplitude versus distance curve as we had presented it, with amplitudes almost 100 times less than reported by the Soviet scientists.

I responded by providing the Soviet scientists with six amplitude measurements from *Rainier*, which supported the concept, as shown in Table 5.1.

These amplitudes were actually from classified AEDS stations operated by AFOAT-1, thinly disguised as "experimental stations." Stations from 820 to 1050 km were the Wyoming Tripartite network and the station at 3700 km was the AEDS station near Fairbanks, Alaska. The station reported to be at 1560 km, but actually at 1610 km, was at Fort Sill, Oklahoma. (I cannot explain this discrepancy.) The station at 280 km was at a now-forgotten location.

As the discussion continued, it became apparent that both sides believed that the first-arriving Pn wave faded into background noise near 1,000 km. Soviet and Czechoslovakian seismologists, however, reported strong P-waves through the mantle arriving at about that same distance, decreasing somewhat with distance initially, then rising to a maximum near 2,000 km.

The U.S. views were colored by Gutenberg's (1954) conclusions that the strong mantle P-waves emerged beginning at about 1,700 km, and our actual observations in the Pacific that Pn was the first arrival out to 2,200 km, with strong mantle P waves beginning at about 1,700-1,800 km (see Figure 2 of Carder's 1964 paper).

I don't think the extent of the shadow zone was ever fully agreed at the Conference; instead we finally chose words for our report that danced around the issue. And perhaps the Soviet data were not convincing, even to them, for one of our agreed conclusions on detectability applies "at a distance of approximately 1,000 kilometres, and also at distances of approximately 2,000–3,500 kilometres." Perhaps a shadow zone lies between?

The Dependence of Amplitude on Yield

In my opening talk I mentioned observations indicating that seismic amplitudes from atmospheric near-surface explosions smaller than about 100 kt increased as the 2/3 power of the yield, and as the first power for higher yields. I also reported that the relationship for underground explosions was not known. In Pasechnik's opening talk, he had assumed amplitude increases at the 1/3 power of the yield. A few days later, on the 17th of July, Dr. Hans Bethe presented a new theory on the coupling of explosive energy into seismic waves. It predicted that the amplitude of P-waves would vary as the 2/3 power of the yield. The new theory predicted, for example, that a ten-fold increase in yield would increase the seismic signal about 4.5 times, or a thousand-fold increase in yield would increase the signal 100 times. In magnitude terms, relative to *Rainier* (magnitude 4 1/4), 5 kt would equate to a little more than magnitude 4 1/2 and 20 kt would equate to about magnitude 5. The theory was well received by the entire conference, and the relationship was incorporated implicitly into the agreed seismic conclusions.

In this talk, Bethe also addressed suggestions made by others (within the U.S. Delegation) that firing a shot in rubble or other loose material might reduce the seismic signal. The hypothesis was that such material might dissipate large amounts of energy in the immediate vicinity of the explosion, leaving less energy to be radiated as seismic waves. Bethe described experiments to test this idea, in which small chemical charges were fired while surrounded by granulated salt. He reported that the salt did, in fact, absorb large amounts of energy. However, it also caused the shock wavelength to increase. The net result was that the longer waves (like the seismic waves that are detectable at large distances) were not much decreased. However, he stressed the lack of good experiments and the need for more observations. This was the only substantive discussion of "decoupling" (the term used for methods that reduce the amount of an explosion's energy that goes into seismic waves) at the conference.

In spite of the Soviet acceptance of Bethe's amplitude-yield scaling law, and a magnitude estimate of 4 1/4 for *Rainier*, a few days later Pasechnik claimed, as he had earlier, that a 5 kt explosion under *Rainier* conditions would create magnitude 5 signals or larger, and thus be easily detectable by existing stations at great distances. Just how an explosion only three times larger than *Rainier* could produce signals almost six times larger was never explained—magnitude 5 for 5 kt seemed to be a matter of dogma.

We pointed out that underwater explosions near Novaya Zemlya of 6 kt and 10 kt (yields previously given by Sadovsky), had been poorly recorded by conventional seismic stations. Both sides agreed that underwater explosions should produce larger seismic signals than underground explosions, yet even the 10 kt shot had been detected by only a few nearby stations in Northern Europe and by a handful of stations at greater distances. Asked to explain this discrepancy, Pasechnik shrugged it off as due to "bad instruments" at the other stations. (Note: There was some merit to his assertion: both shots had been well-recorded at AEDS stations.)

Deep Seismic Sounding Experiments

Pasechnik also cited experiments in the U.S.S.R. in which TNT explosions of several hundred kilograms had been detected at distances up to several hundred kilometers. He illustrated this with seismograms from groups of closely spaced sensors which recorded signals in the 10 cycle per second range. The implicit argument was that if fractions of a ton can be detected to hundreds of kilometers, surely thousands of tons can be detected to truly great distances.

We were familiar with such experiments, which the Soviets called "Deep Seismic Sounding" experiments. Sir Edward Ballard, a renowned British geophysicist, pointed out their irrelevancy—the explosions were often conducted under water where the coupling of energy into seismic waves greatly exceeded that of an explosion underground. Also, they were fired at carefully selected times of night when there was neither wind, nor human activity to create noise in the ground. These were not representative of the circumstances under which a monitoring system would have to function. Furthermore, the recordings were of high frequency signals, which persist to distances up to a few hundred kilometers, but are virtually eliminated by absorption at the greater distances required for a practical test-detection system.

Seismic Background Noise

There were a number of discussions of seismic noise during both formal and informal meetings, revolving around its influence in limiting the detectability of P-waves. It seemed to us that Soviet seismologists had given little thought to microseismic oscillations in the earth in a context of "noise," indeed, their standard "SK" seismograph had been designed to have such a low magnification that normal short period microseismic activity was invisible on the records at most times. In his opening talk, Pasechnik focused on storm microseisms with "periods of five to ten seconds," that reached "even on continental stations, levels of two to five microns." True enough, but hardly germane to considerations of detecting P-waves from small explosions producing amplitudes of ten millimicrons or less well outside the microseismic band at periods of 0.5 - 1 seconds.

At times, it was not clear that we were always talking about the same phenomena as the Russians. At one point, referring to stations in the interior of Europe and Asia, Pasechnik asserted that "during fifty percent of the time interference will be absent altogether." We never did understand Pasechnik's incredible statement, but we did reach common understanding that short-period seismic noise is: low on massive, crystalline rock inside continents; lowest when there is no wind; highest on islands and near coastlines—properties of noise useful for planning a seismic network.

Discrimination between Earthquakes and Explosions

In my opening talk, I reported briefly on *Rainier*, primarily to mention that it produced transverse horizontal shear waves equivalent in size to those from earthquakes with equivalent P-waves—contrary to expectations from an idealized, spherically symmetrical, explosion source.

Concerning a hypothesized "signature" of blasts, I reported that we were "forced to conclude that we know of no universally proven methods for identifying blasts, although we do know methods that might lead us to suspect such an occurrence. Quite obviously, the confidence with which it may be concluded that a blast may have occurred increases with the size of the blast and with the number of observing stations. Absolute proof of the nuclear nature of a blast cannot be obtained by seismic techniques."

I quoted Gutenberg on the annual number of worldwide seismic events worldwide equal to or larger in magnitude than *Rainier* (6000). In the absence of a positive bomb signature, I outlined our approach in separating the many earthquake signals of no interest from those that may be of significance. For this purpose, we first try to recognize and then eliminate signals from shallow earthquakes. "The feature that appears to us the most useful for this identification process is the direction of first motion in the longitudinal wave group." I went on to indicate that detection of "several strong initial rarefactions is a good indication ... of an earthquake source." I cautioned that the amplitude required for this is much greater than for detection only.

Describing a second physically substantiated method for distinguishing between earthquakes and explosions, I pointed out that, "given an adequate number of stations, a high percentage of earthquakes can be identified" as occurring "at depths unattainable to man," and thus eliminated from further consideration.

During Pasechnik's follow-on report, he recited a list of discriminants, most of which were based on simplified and idealized assumptions about generation and propagation of seismic waves. In his discussion of these, Pasechnik cited what he considered to be identifying characteristics, both of explosions and of earthquakes. Explosions would "always" have compressional first motions; they would produce waves with periods one-third of those from earthquakes at the same distance and magnitude; the ratio of shear waves to P-waves would be several times smaller than for earthquakes; and transverse components of shear waves would be near zero. Pasechnik's assertions were not backed by evidence, and none were characteristic of our observations of *Rainier*, although we were confident that the compressional first motion of P was there, but often obscured by noise. Exclusion of earthquakes, according to Pasechnik's ideas, would be based on their focal depth; the occurrence of aftershocks; the existence of Love waves, and the existence of wave periods greater than 5-6 seconds, among other characteristics. Although depth of focus, and the occurrence of aftershocks under some circumstances, could certainly become the basis for valid discriminants, the remainder were conjectural, based on overly simple theory; and some were contradicted by observational data.

In response, Dr. Frank Press, an eminent theoretician as well as an observational seismologist, addressed the subject with the following criticisms:

In order to make the mathematics of theoretical treatments in seismology tractable, the theorist must make such simplifying assumptions about the earth as to cast doubt on their applicability to the subject of this conference. Indeed, the biggest wave [short period shear waves] on some of the *Rainier* seismograms should have been theoretically absent under the conditions of isotropy and homogeneity assumed.

Press went on to draw attention to a recent paper by Dr. Pietro Caloi (1958) who reported on earthquakes in Italy that produced first motions radially outward in all directions where observations had been made (as explosions theoretically do). Press stressed that we must base our conclusions on experimental data, citing other examples where seismograms of real events in the real earth disproved our overly simplified theories. He concluded that

we must consider only what is experimentally realizable at this time. Interpretations must be based on P-wave data alone, recorded at many stations with such clarity as to reveal the direction of first motion.

And indeed, this became the U.S. delegation's guiding principle in our ultimate design of a detection system.

Seismograph Response Characteristics

Dr. Ivan P. Pasechnik, who within a day or two had established himself as the Soviet delegation's leading expert on seismic detection of atomic explosions, argued that underground explosions can be easily recognized by a characteristic waveform. To support this point, he showed numerous seismograms, mostly from high yield surface or low atmospheric explosions. Most were recorded by standard Soviet broadband vertical component Kirnos seismometers (after its inventor, D.P.Kirnos), identified as an "SVK" instrument.[19] Pasechnik used these seismograms to illustrate the existence of a rather simple P-wave pulse, similar in shape from shot-to-shot at different distances when recorded by this instrument. We were familiar with such simple pulses, characteristic of high yield surface bursts as recorded by relatively long period seismometers, but thought these had little to do with our real problem—low yield underground tests, undetectable on such seismographs.

For smaller explosions Pasechnik noted that it was necessary to filter towards shorter periods, although he argued that U.S. instruments used to record *Rainier* had distorted the waveforms. To demonstrate this point, he showed seismograms recording P-waves from four multi-megaton U.S. tests during Operation Castle. For each event he presented side-by-side comparisons of signals from SVK and "VSX" seismographs as well as response curves for both. The SVK seismometer was described as having a natural period of 12 1/2 seconds and it drove a galvanometer with a period of 1.2 seconds. Its magnification was about 1,000 within the period range from 0.2 to 12 seconds, limited by microseismic noise. The VSX seismometer was described as an experimental seismograph having a period of 0.6 seconds, which drove a galvanometer of 0.2 seconds period. It was amplified by some form of photo-electric amplifier, and severely filtered by feedback to reject waves at periods greater than about 0.7 seconds. The result was a sharply tuned system peaking at about 0.3 seconds and achieving a magnification of about 30,000.

The broadband SVK recordings of the four 6.9 - 15 Mt explosions began with a simple compressional pulse followed by smaller oscillations for about a minute. The narrow-band VSX recordings started with small oscillatory waves building up to a maximum two or three times larger than the beginning only after four or five cycles. Pasechnik concluded: "In comparing the records, it will be seen that equipment using narrow bands give a rather serious distortion to the signals."

We had little experience with sharply filtered seismic data at that time, in part because suitable low-noise amplifiers had been available for only a few years, and in part because the filters available to us for operation at the extremely low seismic frequencies had undesirable side-lobes and non-linearities. We relied, as did most seismologists worldwide, almost entirely on the natural mechanical characteristics of the seismometer and galvanometer to shape the response curve so that intermediate period microseismic noise was rejected. Our Benioff short period seismographs were peaked at about 0.3 seconds, similar to Pasechnik's VSX, but they were not severely filtered at longer periods. The recording I had presented showing P-waves from a 1Mt surface burst, although not quite as simple as Pasechnik's SVK recordings, did not show the weak beginning or the complexity of his VSX recordings (refer to Figure 5.4). Even more important from our point of view, the direction of first motion was far more clearly revealed than on any of Pasechnik's recordings of much larger explosions.

In reality, Pasechnik had only shown that his short-period, highly filtered system was not well designed to detect the short period P-waves, even though it discriminated against longer period microseisms. I cannot explain why we did not point this out more forcefully at the time. Probably, we did not believe that Pasechnik's unsuccessful experiments were particularly relevant in view of the success of the short-period Benioff—

well established over many years of operation—and we did not want to seem unduly offensive in pointing this out.

I should explain here that, like the familiar high-fidelity home music system, a flat response from the lowest to the highest frequency produces the most faithful reproduction of the original signal—either sound waves or seismic waves. But if there is also a loud noise present, say the low frequency hum from 60 cycle electric power cross-feed, it may be necessary to filter out the lower frequencies to hear any music at all. In the seismic case there is just such a loud hum, called "microseisms," or sometimes "storm microseisms." These originate from ocean waves that induce energy into the ground near shorelines, from where they may send seismic waves deep into the interior of continents. To detect (hear) weak P-waves with periods of about one-half to one second it is necessary to filter out the low frequency microseisms (usually at four to ten second period). Pasechnik's flat seismometer response curves would give the purest view of P-waves, if they were from explosions large enough to create signals visible above the microseismic noise, but at the cost of failure to detect P-waves from smaller explosions. Our U.S. instruments were designed to detect small P-waves—like those from *Rainier*—at the greatest possible distances, even if at the cost of a small amount of distortion. None of the U.S. team believed that detection and identification of large explosions would be a problem.

Long-period Seismographs

We considered the Soviet SVK and SGK seismographs to be essentially equivalent to long period instruments, responding predominantly to waves having periods on the order of 10 seconds. But Dr. Jack Oliver proposed to the conference a more modern idea of long period instruments. He reported on experience in recording surface waves at periods greater than 15 seconds from earthquakes and large explosions in the Pacific region, and at somewhat shorter periods from Nevada. He cited both similarities and differences between both types of seismic sources, and then described the unusual signals generated by *Wigwam*, the deep underwater explosion in the Pacific. He was cautious about the contribution of long period instruments "in the present stage of development of seismic techniques for detecting nuclear explosions." Nevertheless, he proposed that such instruments should be included in the control system "because of the valuable supplemental information which the long period waves can, on occasion, provide." His recommendation was accepted by the Conference. After many future underground explosions provided the necessary seismic data, the appropriateness of Oliver's insightful recommendation became well established.

Curiously, Pasechnik was unaware of the *Wigwam* explosion, which must have been recorded at numerous stations in the USSR. He asked Oliver for, and received, the location and time of the explosion. Possibly the unique reverberations seen at many U.S. stations were not present, or were unnoticed, on Soviet seismograms, and the explosion had not been identified as such by Soviet seismologists, even though the event was in an aseismic area. This further reinforced our belief that the Soviet seismologists were academic researchers, and not part of an ongoing operational monitoring system.

Seismic Equipment for Control Posts

One of the most troublesome losses of technical content was associated with the agreed description of seismometers to be used in arrays. On 30 July 1958 Dr. Robert Bacher, one of our three delegation leaders, undertook a comprehensive description of the proposed "control posts" (treaty-monitoring stations). He included a description of each type of sensor for monitoring acoustic, seismic, EMP, etc., signals, all to be collocated at each site. We knew that the optimal site conditions for each technique might not be compatible—for example, for the best acoustic reception the terrain should preferably be flat, whereas the best seismic reception was on massive, hard rock like granite, typically found in mountainous terrain —but the Soviets were sensitive about the number of places where foreign operators might be present, so collocation was accepted to minimize the number of stations.

For the seismic system Bacher proposed three types of seismographs, described in general terms in his talk, and in more detail in tables (published with the next day's transcript). The table on seismographs included descriptions of short-period, long-period, and wide-band instruments. The wide-band instrument was intended to accommodate the Soviet's SVK type instrument—so strongly advocated by Pasechnik—while the long-period was proposed by us because of its (unproven) potential to assist with discrimination between earthquakes and explosions, and the short-period instrument was proposed by us to maximize detection of P-waves from low yield tests. It was common practice to characterize seismographs by defining the natural periods of the seismometer and galvanometer, as Pasechnik had done in describing his SVK, (12.5 seconds and 1.2 seconds, respectively) along with damping and coupling coefficients. In Bacher's table the short period seismographs were characterized by "period range 0.2 to 1 second," and "maximum magnification 1,000,000." They were described as electrodynamic or variable reluctance seismometers, and we proposed deploying them in arrays of ten sensors. It seemed clear to us that we were describing Benioff or similar seismographs with seismometer and galvanometer periods of 1.0 and 0.2 seconds or their equivalent, designed to have maximum sensitivity to short period P-waves, as had been discussed during earlier meetings.

Unfortunately, in his more generalized talk, Bacher had used the words "These should have a maximum magnification of approximately one million at 1 cycle per second and should have sufficient bandwidth so that essential signal character is maintained." The word "maximum" in his statement was intended to convey the sense of a magnification "up to" or "no more than" one million at a period of one second—the standard period we used for calibration of short-period seismometers—not that the peak magnification was at one second. The latter would have been in clear conflict with the range of periods (0.2-1.0) specified in his table.

At an informal meeting on 31 July, a Russian draft conclusion on equipment was considered, and translators attempted to reconcile the Russian and English versions. When the English version emerged from the translators, I noted that the description of the short-period instrument for the arrays had dropped the numbers describing the range of periods to be covered, and instead had only quoted Bacher's wording exactly from the verbatim record of his talk. I pointed out that, having dropped all mention of the seismograph's range, 0.2 to 1 second, the wording could be read to imply a *peak* in the response curve at one second, similar to the Soviet SVK-M instrument—and not like the instrument we had proposed. Neither had the Soviets proposed any alternative seismograph for the arrays, so this appeared to be an artifact of the translation process—or the Soviet penchant for eliminating numbers—not a scientific proposal. I suggested an amendment to clarify our intent, but it was late in the day and I was overruled on the ground that the "sufficient bandwidth" clause covered my concern—a serious mistake, as we were to learn a few months later. The conclusions on seismic equipment were adopted the next day in the informal meeting of 1 August.

Seismicity Estimates

Both sides accepted Gutenberg and Richter's (1954) "Seismicity of the Earth" as the basic source—indeed the only authoritative source—for estimating the numbers and worldwide distribution of seismic events with which the control system would have to contend. The authors had compiled a catalog of earthquakes from numerous sources, and recalculated their epicentral coordinates, depths, and magnitudes when the necessary data were available.

It had long been known that small (low magnitude) earthquakes are far more numerous than large earthquakes, but Gutenberg and Richter's analysis quantified this, and showed that the numbers of shallow earthquakes increased exponentially as their magnitudes decreased. Their relationship indicated that magnitude 7 earthquakes were about 8 times more numerous than magnitude 8 events, and magnitude 6 about 8 times more numerous than magnitude 7. For magnitudes below 6, quantitative data were available only from California and New Zealand. When analyzed in a similar manner, these smaller earthquakes from the two regions seemed to follow the same exponential rate of increase in numbers as magnitude decreased. Under the assumption that this rate applied worldwide, Gutenberg and Richter estimated the number of shallow earthquakes that might occur worldwide in various magnitude ranges down to magnitude 2.5. It was a highly uncertain procedure, as the authors recognized, but we and our Soviet counterparts had little choice but to use their estimates.

Inevitably, there were differences in the estimated number of earthquakes equal in size or larger than a given magnitude, or corresponding explosive yield. The Soviets estimated 7,500 earthquakes a year larger than the seismic waves from a one kt explosion; our estimate was 10,000 per year. The 7,500 figure came almost directly from the above referenced 1954 book, while the 10,000 figure came from a later publication (presumably updated) by Gutenberg (1956). Either number was well within the probable error range of the other.

The numbers used by the U.S. Experts as taken from Gutenberg's "TABLE IV" are shown in Table 5.2. The table indicates that we should expect several thousands of earthquakes per year equivalent in size to *Rainier*.

The critical factor was not the gross number of earthquakes to be analyzed by the control system, but rather how many could be identified as natural events. We agreed with the Soviets that events under the ocean could be considered to be identified if no hydroacoustic signals were detected. A study by Norman Haskell and myself of epicenters given in the "Seismicity of the Earth," showed that almost half of all shallow earthquakes could be expected to take place where hydroacoustic methods would assist. The other half, almost 5,000, would need to be identified by first motion or other methods if one kiloton was to be the objective.

Seismic Coupling

The Soviet scientists stated that *Rainier* had been fired in a material that produced exceedingly low "coupling" of the explosive energy into seismic waves. We knew that the amplitudes from the explosions of the same size, but in different materials, could be quite different. We knew that explosions in water created the largest, most detectable, seismic waves, but had only rudimentary knowledge of how the coupling varied among different rock types. Our only standard for underground nuclear explosions was *Rainier*, and we were careful to talk about explosions "under *Rainier* coupling conditions." Sadovsky, at one point, described this as like firing the explosion in "eiderdown," thus strongly muffling the signals. In the end, both sides agreed to a conclusion that *Rainier* conditions were "unfavorable...but even worse conditions are possible." This conclusion was confirmed several years later when the U.S. testing program included shots in a variety of differing rock types.

SIX

The Conference of Experts
The Control System

Having completed reports on the various methods for detecting nuclear explosions, and the equipment to be used for detecting them, the Experts turned to the remaining items on their agenda, namely, description of a system combining the various detention methods, and preparing a report to governments providing conclusions and suggestions for verifying compliance with an agreement or cessation.

Control System Tradeoffs

Design principles for the control system were not developed by the Experts in any coherent way, but important system parameters had been discussed, at least implicitly. Perhaps the most important was the yield to be detected by the system. U.S. experts had no guidance from U.S. policy levels on this matter, but discussions in the Conference often focused on one kiloton, a number that might sound small to political ears. A related parameter was the probability of detection. If the objective was to *detect* any violation then the probability should be high, perhaps 90%; if the objective was only to *deter* cheating, then 50% or 30% probability might suffice.

The number of control posts required to produce data on an event of interest was also an important parameter. Detection at four seismic stations might be enough to determine a crude location, but would be insufficient for identification in most cases; and the location accuracy would be too poor for successful on-site inspection. These parameters might well be different for explosions in different environments, i.e., in the atmosphere, underwater, or underground.

There were painful tradeoffs among these parameters, especially for the Soviets. More control posts on their territory would reduce the number of unidentified seismic events, and thus reduce the number of on-site inspections — but at the price of more places where foreign operators would be stationed. A higher yield objective could be monitored with less foreign intrusion on Soviet territory, but then the U.S. may have had more experience with testing at low yields, and thus might have greater knowledge of how to test undetected at low yield.

The Soviet System

The Soviet delegation's leader, Yevgeni Fedorov, took the initiative on 30 July. He proposed "the simplest possible control system," arguing that all nuclear tests to date had been detected, even without a dedicated control system. To support his point, he tabled a list purporting to give dates and times of 32 nuclear tests conducted by the U.S. in the Pacific between 28 April and 26 July. He claimed that these had all been detected at ranges of 5,000 km or more, and noted that only 14 tests had been announced by the U.S. during that time, so that detecting the others demonstrated the adequacy of existing means.

Two days later Fisk reported back to the Conference that no nuclear tests had taken place at the Pacific proving grounds on two of the listed dates. He requested that Fedorov present the supporting data as a potentially valuable contribution to the conference, citing the need for the control system to avoid false alarms.

Fedorov declined to present the data. No doubt a number of the explosions had been detected by the methods being discussed at the conference — after all, 15 of the tests in this interval had been larger than 200 kt. But very likely the questionable data had actually come from the Soviet spy "trawlers" that lurked around the Pacific proving ground during test detections. The U.S. had made radio broadcasts of "count-downs" at both times, but no nuclear explosion resulted in either case. Far from proving Fedorov's point, we were made even more skeptical of Soviet claims.

Fedorov's proposed network consisted of 100–110 control posts, most of them to be deployed on islands or in coastal regions — places where both Soviet and western Experts agreed that seismic background noise would be high, thus limiting the capability of stations placed there. He claimed the system would be effective for one-kiloton explosions, or smaller. He concluded that "the vast majority" of seismic events would be identified by this system. An international control commission could dispatch a visiting group to the site of a "likely explosion" when "it might become suspicious regarding the occurrence of a nuclear explosion." Such occurrences would be rare, Fedorov predicted, "because the overwhelming majority of suspected cases can arise only as the result of the inaccurate evaluation of earthquakes as a possible underground explosion." The dilemma, of course, posed by this circular statement, lies in deciding which of these "suspected cases" actually resulted from an "inaccurate evaluation" of the control system's data. That was to be the main role of on-site inspection in the majority of cases, in our view.

As was brought out by subsequent discussions the Soviets had placed the primary burden of monitoring underground events on existing seismic stations. In practice, this would mean that monitoring the USSR would depend on data from stations operated by Soviet seismologists. This was unacceptable to us for a host of reasons, including the unsuitability of the Soviet seismometers for detecting low magnitude earthquakes and explosions, and the ease with which such data could be deliberately "spoofed." This latter point may perhaps be best illustrated by considering the first motion of P-waves—critical to discrimination at that time. The recorded motion could be reversed by simply switching two wires connecting the seismometer to its recorder, making (upward) compressional motion appear to be (downward) rarefactional motion. While we of the Western delegations had no objection to data from existing stations—in fact, welcomed them for some uses—we believed that the control posts, operated by a reliable international staff, must be capable of providing all essential, critical information by themselves.

The U.S. Response

A response by us was obviously needed. We knew of no unique way to design a detection network, but we did understand some useful design principles, *e.g.*, geological conditions and desirable distances from noise sources to obtain quiet sites capable of detecting small signals. And we knew how to estimate the detection capabilities of hypothetical station networks using signal and noise amplitudes and their variabilities to calculate signal-to-noise ratios. My notes from the Conference refer to this as the "Haskell Method," and I believe it was Norman Haskell who worked out some of the key statistical aspects of the calculations.

The actual calculations were carried out by Harold Brown and Richard Latter. Brown and Latter undertook the task of defining a network capable of detecting, locating and identifying 90% of earthquakes equivalent in size to events of 1 kt detonated in continental regions under *Rainier* coupling conditions. Both brilliant physicists, I have no doubt that they improved the method in the process.

Jack Oliver and I provided the geophysical inputs. We selected sites having favorable geology, and estimated their probable noise levels based on experience with stations at analogous sites. The signal model was based on the curve shown in Figure 5.5, adjusted to the *Rainier* amplitudes listed in Table 5.1, and scaled to other yields by the Bethe 2/3 power law. Starting with a network similar to that described by Fedorov, which our calculations had indicated would have only about a 5% probability of detecting and identifying 1 kt, we added stations, primarily at inland places where quiet sites could be expected, until the calculated 1 kt capability was achieved. The network required 650 stations! Concerned that presenting this to the Soviet scientists might evoke an extremely negative response, the matter was referred back to Washington for guidance. The guidance that returned was, in effect, "if that's what it takes to identify 1 kt worldwide, go ahead and present it." Shortly thereafter, the study was presented to the Conference by Harold Brown along with a description of the method and the data on which it was based. Dr. Fisk tried to make it clear that we were not *proposing* a 650-station network, but only describing what it would take to identify 90% of events as small as 1 kt.

Although the Soviet scientists made it clear that such a large network was unacceptable, they did not challenge the method of analysis. However, they did question its dependence on the use of first motion as the only basis for identification of shallow earthquakes, and they argued that we had not included the contribution of the existing network of seismic stations. We didn't believe that existing seismic stations would contribute much to detection capability, nor could they be relied on for first motion, for reasons previously explained; and the Soviet scientists were unable to describe any other effective basis for identification of shallow events.

The use of the existing seismic stations seemed to be a major point for the Soviets, and they hammered away on this theme for several days. To us, it seemed obvious that dependence on such stations was tantamount to self-policing; but as scientists, we were reluctant to express distrust of fellow Soviet scientists (who were, however, employees of the Soviet Government). Eventually, Jim Fisk frankly explained the ease with which the direction of first motion could be made unreliable. He also pointed out that the Soviet proposal would either shift the verification burden to systems outside of the control system, or the hundreds of existing stations would have to be incorporated into the system—effectively making the system even larger than the 650 station network described by Dr. Brown.

A British Compromise Proposal

A few days later Sir William Penney proposed that the Conference set a threshold for the system at 5 kt, in view of the difficulty of achieving a 1 kt threshold. He also rejected the idea of splitting the difference between the Soviet's number of control posts and the number mentioned by the U.S., which would have resulted in a cumbersome system of 380 control posts, while still not having good capability at 1 kt. A compromise system, designed by U.S. physicists and seismologists working with Penney, was presented by Penney to the Conference on 5 August. It would consist of about 170 stations on land, and 10 on ships (for acoustic and EMP detection, and for fallout collection). The system would still have some capability at lower yields, particularly detection capability; and he suggested that events in aseismic (non-seismic) regions down to the 1 kt level should automatically be inspected. He went on to report that about 60 stations, mostly located in inland aseismic regions would be at "very quiet" sites, 60 others would be at "quiet" sites, and the remaining 50 in coastal regions would be at "noisy" sites. He reported that this system would have a 90% identification capability at 5 kt over much of the continental regions of the world, but cautioned that the capability would be less in seismically active island chains.

Three systems were now on the table, the Soviet system with little capability at 1 kt, and the large and unwieldy system described by the U.S., but with good capability at 1 kt. Fisk argued that each should be described in a final report, along with a discussion of its capabilities and limitations — indeed this was what he had been requested to do. Fedorov would not agree. He did, however, agree to "compromise" and report on the Penney system. The best we could manage to do in an agreed report was to have qualitative language included, to describe the effect of changing some of the characteristics of the system. Station locations, as envisioned by the U.S. Experts, and which were used for estimating capabilities of the 170-station system, are shown in Figure 6.2.

The Experts' Report

A final report was concluded on 20 August 1958. It contained the agreed conclusions on the applicability of each of the several detection methods, and the description of the equipment to be installed in the control posts. The report then outlined characteristics of a hypothetical worldwide network of 160-170 "control posts" and ten ships, under the direction of an "international control organ," and equipped with the appropriate apparatus to detect subsurface and low atmospheric nuclear tests. Aircraft sampling for radioactive debris was included, as were provisions for on-site inspection. It also outlined the estimated capabilities and limitations of the system.

Evasive testing methods were not considered in the Experts' report, although it admits to difficulties when the explosion is carefully concealed. Specification of the means for monitoring high altitude explosions was left as unfinished business.

Selected key findings applicable to underground explosions were:

1) When nuclear explosions occur under the ground or under the water, longitudinal, transverse and surface waves are

formed and get propagated to great distances. The first longitudinal wave is the most important, both for detecting an explosion and for determining the place of the explosion, and also for distinguishing an earthquake from explosions. Transverse and surface waves also help to define the nature of a seismic perturbation.

2) Longitudinal seismic waves caused by underground nuclear explosions set off under conditions analogous to those in which the *Rainier* shot occurred can be detected, and the direction of first motion of the longitudinal wave can be determined at a distance of approximately 1,000 kilometres, and also at distances of approximately 2,000-3,500 kilometres at sites which are considerably more quiet than the average for:

   a. explosions of the order of one kiloton recorded during periods of favorable noise conditions.
   b. explosions of the order of five kilotons recorded during periods of unfavorable noise conditions.

   It must be noted that all seismic stations situated at thousands of kilometres from one another cannot have an identically high or identically low level of background at one and the same time.

3) Conditions for detection and identification of underwater explosions set off in shallow water but at a sufficient depth, are considerably more favorable than conditions for detecting underground explosions.

4) Control posts carrying out seismic observations should be put at sites with a minimal level of microseismic background, such as are possible in internal continental regions. Such stations, when provided with arrays of seismographs, can insure the obtaining of the data indicated above. However, at stations that are in unfavorable regions such as coastal and island regions the noise level will be higher than at quiet stations inside continents. In these cases for detection and determination of the sign of first motion, the energy of the explosion must increase in the ratio of the power of 3/2 with respect to the increase of background level. This is in part compensated for by the fact that quiet stations inside continents will register more powerful explosions at distances of from 2,000 to 3,500 kilometres. Bursts with an energy of 5 kilotons and more will be detected by quiet stations placed at the distances named.

5) The majority of earthquakes can be distinguished from explosions with a high degree of reliability if the direction of first

motion of the longitudinal wave is clearly registered at 5 or more seismic stations on various bearings from the epicenter...

...With modern methods and making use of the data of several surrounding seismic stations the area within which an epicenter is localized can be assessed as approximately 100-200 square kilometres.

The agreed "seismic apparatus" was to consist of seismometers with four different response characteristics:

1) Approximately 10 short-period vertical seismographs dispersed over a distance of 1.5-3 kilometres and connected to the recording system by lines of cable. The seismographs should have a maximum magnification of the order of 10^6 at a frequency of 1 c.p.s. and a receiving band adequate to reproduce the characteristic form of the seismic signal.
2) two horizontal seismographs with the parameters indicated in point 1.
3) One three-component installation of long-period seismographs having a broad receiving band and a constant magnification of the order of 10^3-2×10^3 in the period range 1-10 seconds.
4) One three-component installation of seismographs with a narrow receiving band and magnification of the order of 10^4 - 2×10^4 at periods of $T = 25$ seconds.
5) At certain points one three-component installation of long-period seismographs with magnification of the order of 10^4-2×10^4 at periods of $T = 25$ seconds.
6) Auxiliary equipment necessary in order to get precise records of the seismic signal: recording devices, chronometers, power supply units and apparatus for receiving automatic radio-signals giving correct time.

Introducing the system, the report stated:

The conference considers that the problem of detecting and identifying underground explosions is one of the most difficult, and that, to a large extent, it determines the characteristics of the network of control posts.

The spacing between the control posts in continental aseismic areas would be about 1700 kilometers, and in seismic areas about

1000 kilometers. The spacing between the control posts in ocean areas would vary between 2000 and more than 3500 kilometers; the spacing between island control posts in seismic areas would be about 1000 kilometers.

The system would have:

Good probability of recording seismic signals from deep underground nuclear explosions in continents equivalent to 1 kiloton and above.

On identification the report stated:

Along with the observation of signals of possible underground explosions the control posts would record at the same time a considerable number of similar signals from natural earthquakes. Although with the present state of knowledge and techniques, the network of control posts would be unable to distinguish the signals from underground explosions from those of some earthquakes, it could identify as being of natural origin about 90 percent of the continental earthquakes, whose signals are equivalent to 5 kilotons...

It has been estimated on the basis of existing data that the number of earthquakes which would be undistinguishable on the basis of their seismic signals from deep underground nuclear explosions of about 5 kiloton yield could be in continental areas from 20 to 100 a year. Those unidentified events which could be suspected of being nuclear explosions would be inspected...

Concerning on-site inspection, the report stated:

When the control posts detect an event which cannot be identified by the international control organ *and which could be suspected of being a nuclear explosion*, the international control organ can send an inspection group to the site of this event in order to determine whether a nuclear explosion had taken place or not. The group would be provided with equipment and apparatus appropriate to its task in each case. [italics added]

Thus, the technical basis for on-site inspection to aid in nuclear test monitoring was enunciated for the first time in an East-West agreed document. At the time, this was considered by many in the West to be the most important contribution of the Conference of Experts. However, the phrase in italics was symptomatic of a deep difference of view as to the circumstances that call for on-site inspection. To the Western side, virtually any unidentified seismic event could be suspected of being a nuclear explosion. Each would thus be *eligible* for inspection, although few would actually be selected for inspection. The number of inspections would be limited by the available trained personnel and funds, if for no other reason. In the selection process other factors would be considered. For example, seismic events under highly populated areas, or in totally inaccessible regions, or when other circumstances made an explosion highly unlikely, could be exempted from the process.

To the Soviets, the phrase meant that there should be highly positive reasons for suspicion, perhaps such as location on a known test site. Our approach was to employ criteria that would never misidentify an explosion as an earthquake. Their approach was to employ criteria that would limit inspections to as few areas and as few times as possible. This difference was made abundantly clear only after treaty negotiations began, at the end of October 1958.

The Call for Treaty Negotiations

The report of the Conference was released on 21 August 1958, along with an optimistic press release. A key conclusion in the release stated that "it [is] possible, within certain specific limits to detect and identify nuclear explosions, and [the Conference] recommends the use of these methods in a control system."

President Eisenhower reflected this optimism in his press conference later that day, and on 22 August 1958 he proposed that the three nuclear powers negotiate a treaty suspending all nuclear tests permanently. He announced that the United States would suspend all tests for one year commencing with the start of negotiations provided that the other nations do the same, and that the suspension would be extended beyond the year provided that satisfactory progress had been made toward establishing a control system and in other areas of arms control.

A Few Comments on the Report

That unidentified seismic events would be the primary reason for on-site inspection to aid the control system was clear enough in the Experts' report. What was perhaps not so clearly revealed by their conclusions is the reason. Basically, the relative ease of identifying explosions in the atmosphere is explained by the fact that they are detectable by several methods that provide both overlapping and complementary capabilities.

For instance, acoustic signals and radio signals are both generated by atmospheric nuclear explosions and, when both are detected, may uniquely identify the source as an explosion. This provides a way to eliminate false alarms from either method that might be caused by natural events. Furthermore, the applicable detection methods were well understood and had been thoroughly tested on large numbers of atmospheric explosions. But of most importance, radioactive debris is usually also detected, and it provides unequivocal evidence to establish the nuclear nature of the detected event.

By contrast deep underground nuclear explosions are detectable at long range only by seismic waves (as are the earthquakes that may create false alarms). In general, the detected seismic signals from an explosion do not positively identify it as such. Neither can earthquakes always be identified positively. The focus here, of course, as it was for the Experts, is on low magnitude events. The consequences, after going through the identification process, is that there is a residuum of unidentified events, some of which might be explosions. It should also be noted that detection, and especially identification, methods were less well understood since there had been only one underground nuclear explosion prior to the Experts' Conference—the *Rainier* explosion of 1957. Furthermore, even if there is a high degree of certainty that the event is an explosion, there is nothing in the low magnitude seismic signals that proves the *nuclear* nature of a detected event. On-site inspection offers the possibility, at least, of acquiring nuclear proof, and this constitutes one of the underlying objectives of on-site inspection.

Although the Experts recommended the *methods* for test monitoring, they did not advocate the hypothetical *system* as a system designed to meet the specific test ban monitoring objectives of any particular nation represented at the meeting. Such objectives were not discussed and, as far as I know, have not been clearly defined by the U.S. to this day. The report indicated, in a qualitative way, what the effect of increasing or decreasing the number of control posts would be, implicitly leaving it to governments to decide whether greater or lesser capabilities should be negotiated during the political conference to come. However, the control system was recommended for "consideration by governments." When nuclear test ban treaty negotiations began in late October, the Experts' report was accepted as the technical basis for negotiation, and the Experts' control system became, de facto, *the* monitoring system to be discussed.

Within seismological circles the system has commonly been described as one designed to record signals at local or regional distances. In fact, it was designed to record signals predominantly from teleseismic distances, and its capability rests largely on teleseismic signals. This results from the estimate that signals are as large in the zone between 2000 and 3500 kilometers as they are out to about 800 km, and the number of stations in the teleseismic zone would generally be substantially larger than the number within 800 km, since the area included is much greater.

A Comment on Arrays

Advocating seismometer arrays for the detection of P-waves at great distances was a significant departure from normal seismic observatory practice at the time. However, such arrays were necessarily limited in size and number of sensors by the technology of the time. Since signals could only be directly and instantaneously summed electrically, it was necessary to keep the dimensions of the array smaller than one-half of the shortest wavelength to be detected (so that the signals would be nearly in phase) to achieve coherent addition of the signals. Teleseismic signals of interest could be as short as 8-10 km, which led to the Experts' 1.5 -3 km dimensions for the array. Within an area of those dimensions, however, seismometers must be spaced far enough apart that the short period noise would be uncorrelated between sensors to achieve partial noise cancellation. This separation had been found to be on the order of one kilometer. This, in turn, limited the number to about 10 seismometers.

A Comment on Location Accuracy

At the time of the Experts' Conference, there was no generally accepted and rigorous method for assessing location accuracy. Their conclusion, "making use of the data of several surrounding seismic stations the area within which an epicenter is localized can be assessed as approximately 100-200 square kilometers" is far too optimistic as we would later find from additional explosions. Such areas imply calculated locations within 6-8 km of the true location. Subsequent data from large explosions have demonstrated the existence of biases of up to tens of kilometers. Even without systematic biases, modern estimates indicate it would require signals recorded at dozens of stations to achieve such accuracies (see Appendix D).

SEVEN

Hardtack II: New Seismic Data

Operation Hardtack I had commenced 28 April 1958 with a low yield device carried aloft to 26 km by a balloon northeast of Bikini. The series consisted mostly of shots detonated on barges at Bikini and with yields ranging from a few kt to almost 10 Mt. There were also two underwater shots of 8 and 9 kt at depths of 50 m and 150 m respectively. Hardtack I culminated in August while the Conference of Experts was still in session with two rocket-launched 3.8 Mt high altitude shots near Johnston Island (about 1,250 km WSW of Hawaii) producing widespread radio blackouts and spectacular auroral effects seen in both northern and southern regions of the Pacific.

In Geneva, while concluding the Conference of Experts, we were acutely conscious that the planned second phase of Hardtack, to take place at the Nevada test site during late September and October, would include underground nuclear explosions that might provide data that could settle contentious issues still being disputed at the Conference. The Bethe Working Group earlier had seen the need for better (and more) data on underground tests and had urged the President not to accept a moratorium on testing until Hardtack II was completed. Immediately after the Conference of Experts ended, President Eisenhower proposed negotiations with the other two nuclear powers for the suspension of nuclear testing on the basis of the Experts' report. He proposed that negotiations begin 31 October 1958, and that a one-year testing moratorium should begin with the start of the negotiations. Facing this deadline, the NTS staff began conducting safety tests and other low yield shots (<100 t), and rushed to complete preparations for larger underground explosions.

The Seismic Measurement Plan

In AFOAT-1, we exerted all possible effort to prepare for and conduct a program to measure the amplitudes of seismic waves from the underground shots. The work was carried out under the immediate command of Lt. Col. Benjamin Grote and Major John Bush. With the assistance of a contractor (The Geotechnical Corporation), under the leadership of Joseph Whalen and Wayne Helterbran, we were able to equip and install 18 temporary seismographic stations. These stations were disposed at distances extending from 60 km from the planned underground shot points to a distance of slightly more than 4,000 km along a line extending generally eastward from the Nevada Test Site to Arkansas and thence northeastward to Maine. Each operating site and alternate sites had been preselected by a team of geologists and geophysicists who located suitable outcrops of hard rock as far away as possible from obvious sources of manmade seismic noise and arranged with the land owners for operating rights.

It was planned that eight temporary stations would remain at fixed locations to provide a consistent set of data on all shots, and it was hoped that the remaining ten mobile stations could be moved between shots in order to record data at different distances. These mobile teams were, in fact, able to move between the two largest underground shots *Logan* (5 kt) and *Blanca* (19kt).[20] Thus, recordings were made by the temporary teams at 28 different distances from the shot points, and at the array station at Ft. Sill, Oklahoma. As a result, recordings were made at intervals of approximately 100 to 200 km out to 2500 km, and at intervals of approximately 200 to 300 km thereafter to a distance of 4,020 km. In addition to data along the main temporary profile, both *Logan* and *Blanca* were recorded at the three AEDS stations in Wyoming and the station in Alaska. Data from two of these classified stations were released and included in my published paper (Romney, 1959), and data from all four were released to Soviet seismologists slightly later. *Blanca* was also recorded at three AEDS stations in foreign countries; as far as I know these data have not been publicly disclosed.

In addition to *Logan* and *Blanca*, there were three sub-kiloton underground shots: *Tamalpais*, 72 t; *Evans*, 55 t; and *Neptune*, 90 t. Like *Rainier*, all but *Neptune* were named for well-known mountains in the western U.S. and Canada. Although not recorded beyond about 300 km, the sub-kiloton tests proved to be useful for extending our amplitude versus yield curve to yields lower than *Rainier*.

The temporary stations were all equipped with Benioff short-period vertical seismographs and most of them were additionally equipped with two horizontal components that were oriented radially along and transverse to the path from the shot point to the station. Each station was provided with an accurate timing device and means for recording radio time signals broadcast from WWV on a continuous basis. Each seismograph was calibrated under operating conditions by means of a Geotechnical

Corporation ball-lift calibrator, which also verified the polarity of the instrument. In addition, special efforts were made to check both the calibration and the polarity by independent methods at one time or another during the course of the observations. The periods and damping ratios of the seismometers and galvanometers gave the response characteristics as outlined in Professor Robert Bacher's Table presented to the Conference of Experts. Seismometers and galvanometers were loosely coupled and near-critically damped to avoid resonance effects. The magnifications of the instruments were known within 5 to 10 percent. It was the largest well-calibrated seismic experiment since the secret Buster-Jangle Operation in 1951.

Each of the Hardtack underground explosions produced almost identical waveforms at a given station. It was therefore possible to measure the amplitude recorded from each shot at several identifiable points within the wavetrain. The separate measurements were then divided by the corresponding amplitudes recorded from the *Logan* shot, so that all results were normalized to *Logan* amplitude. For example, at the temporary stations, the average ratio of the *Blanca* amplitudes to the *Logan* amplitudes was 2.36. At 12 stations of the University of California and Caltech, the average amplitude ratio on Wood-Anderson seismographs was 2.22. The results from these two sets of observations thus agreed within a few percent.

The larger explosions were also recorded by many stations operated by universities, the Dominion Observatory of Canada, and the U.S. Coast and Geodetic Survey; the latter organization collected seismograms from all of these stations and provided copies for AFOAT-1's use (our secret mission prevented us from directly requesting such data). Stations operated by Caltech and the University of California were especially useful, since a number of them employed well-calibrated Wood-Anderson torsion seismographs that had also recorded the *Rainier* explosion. These stations thus provided a reliable link between amplitudes and magnitudes (M_L) of *Rainier* and those of the Hardtack II shots. Other stations, with widely differing response characteristics, and frequently with uncertain calibrations, provided data useful for studies of travel-times and the direction of first motion in the P-waves, and they gave important observations of other-types of waves generated by the explosions.

The main thrust of our experiment, however, was to measure the amplitudes of the explosions on seismographs with accurately known and consistent response characteristics, identical to those used at our classified AEDS stations. With such data, we could investigate such significant questions as the variations of P-wave amplitude with distance and the dependence of P-wave amplitude on yield. With successful P-wave measurements at teleseismic distances (for this purpose meaning beyond 16°, the range in which m_b was defined by Gutenberg and Richter [1946]) we could, for the first time, directly relate the m_b and M_L scales for events in the magnitude 4-5 range (see the section on magnitudes that follows).

Magnitudes

Early methods of ranking earthquakes according to their size were based mainly on their effects on structures. However, these methods could result in moderate earthquakes near population centers being rated as large, and truly large earthquakes in remote areas being ignored.

The concept of "magnitude" based on the motions of the ground measured by seismographs was invented by Charles Richter (1935) of Caltech. Richter's original scale corrected measured seismic amplitudes to a distance of 100 km. Measurements were made of the seismic waves recorded by standard Wood-Anderson torsion seismographs, and the logarithm of the largest amplitude was taken to represent magnitude. The amplitudes were measured in arbitrary units, selected to make a "zero magnitude" event almost unmeasurable. Thus magnitudes 1, 2, 3, etc. produced ground motions 10, 100, 1,000, etc. times larger than the zero magnitude event.

As originally defined, the magnitude scale is applicable only to measurements at local distances. Working together, Gutenberg and Richter attempted to extend the scale to great distances, based on measurements of long period (~20 second) surface waves. The resulting scale was denoted "M_s." They also attempted to extend the scale to great distance by using measurements of body waves, particularly P-waves. This measurement is commonly denoted "m_b." Gutenberg also defined a "unified magnitude" combining surface wave data with body wave data. He denoted this unified magnitude as "m," but he concluded that all methods had been successfully normalized to m_b. Thus the unified magnitude, m, should be the same as m_b.

Amplitude-Distance Relationship for P-Waves

The P-wave amplitudes as a function of distance are shown in Figure 7.1. The data indicated by the solid circles are from *Blanca* (19 kt). *Logan* data, indicated by open circles, were used to help define this curve by multiplying each observed *Logan* amplitude by 2.36 (as previously noted, this was the average amplitude ratio of *Blanca* to *Logan* at fixed stations).

At distances between about 200 km and 1000 km the amplitudes of P (called Pn in this distance range) were found to be inversely proportional to the cube of the distance, a result consistent with a number of previous earthquake studies. Beyond about 1,000 km, Pn decreased to such small amplitudes that it became undetectable at most stations although there were several observations that seemed to fit a Pn travel-time curve out to about 1,300-1,400 km. However, beginning at 1,100 km other strong impulsive P waves were detected several seconds after the expected time of Pn. Unlike Pn at short distances, and P waves that travel through the mantle at greater distances, these waves often began with weak precursory motions followed by the stronger pulse. We could not be certain that the first motion was actually observed in many cases (see Fig. 7.2). Generally, these late waves had longer periods than Pn, and the travel times were close to Gutenberg's 1953 travel-time curve.

The abrupt change in the amplitude-distance relationship, the longer periods, and the break in the travel-time curve all indicate the arrival of waves along a new path. Similar observations were noted by Gutenberg and discussed extensively by him in a paper on the structure of the Earth's mantle (Gutenberg, 1954). He concluded that "the first recorded waves . . . are very small," (see my precursors in the paragraph above) and he concluded they were probably "diffracted waves." Uncertainty about the nature of P-waves between about 1,000 km and 1,500-1,700 km led Gutenberg to define his m_b scale starting at 16 degrees from the epicenter and beyond, (about 1750 km), where it was clear that the arrival was a true mantle P-wave.

Beyond 2,000 km the amplitudes fall off about as expected from earthquake studies, except that there appeared to be a low-amplitude region between about 3,000 and 3,500 km, as we had noted from shots at the Pacific test sites. *Logan* was not detected at 3,000 and 3,500 km and *Blanca* was not detected at 3,300 km. In each case, measurements of microseismic noise showed that the P-wave amplitudes must be less than 5 to 10 millimicrons (see Appendix E).

Amplitude-Yield Relationship for P-waves

Only *Blanca, Logan,* and *Rainier* produced measurable P waves at distance beyond about 300 km. Amplitude ratios on the smaller shots were therefore less well determined. The results of these observations are shown in Figure 7.3. It may be seen that the points fit an approximate first-power relationship between amplitude and yield, rather than the 2/3 power relationship adopted by the Conference of Experts based on Professor Hans Bethe's theory. Well, remembering Perry Byerly's advice that "the data are the important thing" and "theory will follow data," our finding was quickly imparted to colleagues concerned with nuclear test detection. And sure enough, a theoretical analysis of the reasons for this first-power relationship was quickly developed and circulated within official circles, by Latter, Martinelli, and Teller (1959).

For a comparison, the amplitudes recorded on the Wood-Anderson Torsion seismographs at Tinemaha are plotted on the same graph. These measurements are of the maximum amplitudes in the shear waves, rather than in the P-wave group. It may be seen that the Tinemaha shear-wave amplitudes have about the same dependence on yield as the amplitudes of P waves.

For reasons that are still not understood, the amplitudes from *Evans* were slightly less than one-tenth of those from *Tamalpais*, although the two shots released similar amounts of energy (55 and 72 tons, respectively). This phenomenon was termed "The *Evans* Mystery" at the time; it cautioned us that the seismic waves from underground explosions may be highly sensitive to small changes in rock characteristics, chamber size, or other local parameters.

Magnitude Versus Yield

Magnitudes of Hardtack II explosions were first determined for all shots recorded by Wood-Anderson torsion seismographs according to the original method devised by Richter. Data were available from ten stations, giving an average local magnitude, M_L for *Blanca*, of 4.8 ± 0.4,[21] and for *Logan*, 4.4 ± 0.4 (see Appendix E).

Body wave magnitudes, m_b were next computed for *Blanca* and *Logan* from eight and six stations, respectively, at greater distances. In the computations the procedures and the table given by Gutenberg and Richter (1956) were used. The results are: for *Blanca* $m_b = 4.8 \pm 0.4$; and for *Logan* $m_b = 4.4 \pm 0.5$. Thus magnitudes estimated by m_b agree with the estimates based on M_L.

Gutenberg and Richter had cautioned that "the relation of M_L to m is not yet on a definitive basis"; but they gave an equation that indicated that the two magnitudes may differ by about 1/2 units at the observed *Logan* and *Blanca* magnitudes. To test this relationship they showed a plot of m_b versus M_L for 15 large events (Figure 7.4), and commented that their equation "is not inconsistent with the plotted data." As I observed at the time, neither is $M_L = m_b$ *inconsistent* with the plotted data. Gutenberg and Richter had attempted to define the newer m_b scale so that it agreed with the M_L scale; our data indicated that they may have succeeded after all.

Since it appeared that the magnitudes by both methods were a homogeneous set, having the same mean and essentially the same standard deviation, all measurements were combined to give a magnitude of 4.8 for *Blanca* and 4.4 for *Logan*. I hasten to add here that there is no compelling physical reason to believe that the relationship between M_L and m_b that we observed for explosions also applies for earthquakes. It can be argued on physical grounds that M_L, measured on the maximum of the short period shear waves, should be greater for earthquakes than for explosions having equivalent P-wave amplitudes (and m_b values).

Only seven stations with torsion seismographs recorded what were originally considered to be accurately measurable magnitudes for *Rainier*; these gave a local magnitude of 4 1/4 ± 0.2. However, there is another way to estimate the magnitude of *Rainier*. We found that, on average, *Rainier* magnitudes at each station are 0.75 smaller than *Blanca*'s and 0.35 smaller than *Logan*'s. Accepting that magnitudes of *Logan* and *Blanca* are better determined (more stations, larger signals) both differences indicate that *Rainier's* magnitude was actually about 4.05, rather than 4.25 as previously reported. Implicitly, the missing data from three stations, Woody, Barrett, and Riverside (all in California) would have given smaller magnitudes than data from the seven stations included in our initial estimates.

I have gone into this in some detail because it later became an issue between us and the Russians. It is also an illustration of *magnitude bias*, which has led seismologists to a number of erroneous conclusions over the years. The underlying cause of this bias is that seismic amplitudes from an explosion (or earthquake) vary widely from place to place, even at the same distance. Thus, for events in the magnitude 4-5 range, the smaller signals may be obscured by noise, so that the seismologist may see and measure only the signals that are larger than average, leading to a biased magnitude estimate. This had occurred in the case of *Rainier*: when we later measured magnitudes from recordings initially rejected as too weak to be reliable at Woody, Barrett and Riverside, our best estimates were 3.6, 3.3 and 3.8, respectively, confirming that the best estimate for *Rainier* was 4.05, not 4.25.

The magnitudes of the subkiloton shots may also be estimated from their amplitudes relative to the larger shots. Combining the magnitude information with the known yields results in the relationship:

$$m = 3.65 + \log Y,$$

where Y is the yield in kilotons.

Identification Characteristics

The Hardtack II explosions produced data with strong implications for methods of discriminating between earthquakes and explosions. Concerning the direction of first motion—the discriminant proposed by the Expert's Conference—I reported (Romney, 1959) that:

It may be seen that the first motion is recorded as compressional at some stations and rarefactional at others. There seems to be no systematic dependence upon the distance, with the obvious exception that the first wave is strong and compressional at small distances (less than about 700 km). At distances greater than 1100 km and less than 2650 km, it is not known with certainty whether the first motion was observed at any station on either *Logan* or *Blanca*. At some stations, (for example, the station at 2,300 km) there is apparently clear rarefactional first motion in spite of a signal-to-noise ratio of at least ten.

Both travel times and amplitudes of P-waves beyond 1,100 km suggested that they were normal P-waves refracted through the earth's upper mantle. However, as previously noted, there were indications of a precursor at a number of stations at distances less than about 2,500 km. Figure 7.2 illustrates this phenomenon, where it may be seen that, as the signal-to-noise ratio increases, weaker and earlier waves emerge above the noise. These precursors were not understood in terms of classical ray theory. Their effect, however, was to make the observed first motion unreliable in this zone.

At greater and lesser distances, we believed that we could make reliable picks of the first arrival and its direction of motion at many of the temporary stations. These were used to construct an amplitude versus distance curve for the first motion of P, applicable outside the zone of precursory arrivals (see Figure 7.5). Comparing this to the peak of the P-waves, we also noted that the amplitude of the first motion decreased with distance at a more rapid rate than did the main P-pulse.

An additional, and disappointing, characteristic of first motion from *Logan* and *Blanca* was that it was smaller than expected *relative* to the half-cycle that followed. Our expectations had been strongly colored by observations of first motion from near-surface air bursts. For example, the Hardtack II explosion, *Socorro,* a 6 kt shot on a 60 m tower, produced first compressional motions more than 1/2 the amplitude of the following, rarefactional, motion at most stations. For *Logan* and *Blanca,* on the other hand, the first motion was typically less than one quarter of the amplitude of the following motion. Since these observations were made only days apart, at the same stations, and on the same instruments, it was clear that this was a near-source effect. We initially postulated that the reflection from the surface above the underground explosions interfered with and blunted the first motion. Werth et. al. (1962) later showed by modeling that this was not the case; rather, the reflections probably enhanced the following rarefactional motion. The implications, however, were clear whatever the cause: the ratio of the peak signal to the noise would need to be substantially greater than we had assumed at the Experts' Conference, to be certain that we didn't misread the true direction of the first motion. This, in turn, meant that our principal identification criterion was less effective than we had earlier estimated.

Horizontally polarized surface waves (known as "Love waves" after the mathematician A.E. H. Love) were generated by the larger explosions. I reported "...there are clear indications of Love waves at Berkeley and Palisades; and at Resolute Bay the surface wavetrain consists predominately of Love waves of 10- to 15-sec period. These latter waves were completely unexpected from blasts,..." Thus, the mere existence of Love waves is not proof of an earthquake, as had been thought previously.

USC&GS Independent Analysis

The U.S. Coast and Geodetic Survey also conducted their own independent analysis of the seismograms they had collected from numerous unclassified conventional stations. Their results qualitatively confirmed our findings on the detectability of P-waves and the direction of first motion. Their results are summarized as follows in a letter of 12 November 1958 to the AEC's General Alfred Starbird from Rear Admiral Charles Pierce of the USC&GS:

This is a preliminary report of seismograph stations in this country and a few foreign countries that have detected the larger underground explosions at the Nevada Test Site during October 1958.

In preparing the material appearing in Table I a telegraphic inquiry was sent to a restricted number of stations to determine the relative range over which the explosions might have been recorded. For *Tamalpais* the list of stations was restricted primarily to the western United States and a few stations in the East, Alaska, and Canada. Positive reports were received from 13 stations. For *Logan* the list was increased to include more stations in the East. Positive reports were received from 38 stations, the greatest distance being 2310 miles to College, Alaska. For *Evans* an exhaustive canvass was made of nearly all the United States and many foreign stations. Not one station reported a recording. The same distribution was made for *Blanca*. Many of the United States stations, 33 to date, have reported definite recordings. Reports are incomplete for several networks, for example, California Institute of Technology, University of California, University of Michigan. Several foreign stations with sensitive seismographs have reported recordings of *Blanca*—Huancayo, Peru; Uppsala, Sweden; Matsushiro, Japan.

In Table II we have a preliminary report of stations having sent their *Logan* and *Blanca* seismograms to the Coast and Geodetic Survey. All stations that submitted telegraphic reports of *Logan* and *Blanca* together with a number of others have been asked to loan us their seismograms. In the Table are our readings for P and S and the direction of first motions. In general the impulsive type reading has better definition than the emergent. There are some differences in the types of first motions (compression and dilatation) even for stations recording well-defined phases. This is surprising in view of the fact that theoretically explosions should send compressional waves in all directions from the source.

Several pages of tables (not included here) appended to Admiral Pierce's letter showed that dilatational first motion of P-waves were recorded from *Logan* beginning at Salt Lake City, 572 km. The report also illustrates that a 5 kt shot was not reported worldwide, contrary to Pasechnik's assertions at the Conference of Experts; and even a 19 kt shot was reported from only three conventional stations outside of North America, as well as from three classified AEDS stations equipped with arrays.

Treaty Negotiations Begin

Meanwhile, on 31 October 1958 "the Conference on the Discontinuance of Nuclear Tests" had convened in Geneva, attended by ambassadors of the three nuclear powers, the U.S., U.K., and USSR. *Blanca*, our most important test, had been detonated only on the previous day. Photographic records of the several underground shots were assembled at AFOAT-1 as quickly as possible and analyses had commenced.

Waiting impatiently in Geneva for the results of our analysis, my boss, Doyle Northrup, a member of the U.S. delegation, soon summoned me, and I arrived Thanksgiving Day. A day or so later, I briefed Ambassador James J. Wadsworth, U.S. Representative on Disarmament and Deputy U.S. Representative to the U.N., together with Ambassador Sir David Ormsby-Gore, head of the U.K. Delegation and Minister of State for Foreign Affairs. Ambassador Semyon Tsarapkin, who represented the USSR, was not briefed at that time—our results were still classified SECRET.

Even at that early stage, it was evident that the Experts' conclusions were not borne out by our new Hardtack II data. Work started immediately on drafting a report to be tabled at the conference. My stay in Geneva was short and I returned to Washington to complete the analysis of the seismic data.

The Ad Hoc Panel's View

When the new data from Hardtack II had been more completely analyzed, we convened an "Ad Hoc Panel of Seismologists" to review and assess it. The Panel met on 16-19 December 1958 at AFOAT-1 Headquarters, at that time on Telegraph Road south of Alexandria, Va.

I chaired the meetings, and also presented the data and tentative conclusions to the Panel, ably assisted by Major Mark Colvin, a member of my staff. Other Panel members were: Billy G. Brooks, then Chief Seismologist, The Geotechnical Corporation; Professors Perry Byerly, Frank Press, and Jack Oliver, who had participated in the Experts' Conference; Dr. Dean Carder of USC&GS; and Professor James T. Wilson, Chairman, Department of Geology, University of Michigan. Also present at times were Professors Hans A. Bethe and David T. Griggs (University of California at Los Angeles), both of whom made comments and suggestions from time to time as the seismologists reviewed the data and debated its implications for nuclear test monitoring. Dr. Kenneth Street, University of California Radiation Laboratory, and Dr. J. Carson Mark, Los Alamos Scientific Laboratory also attended briefly. These latter four scientists happened to be at AFOAT-1 at the time—for a coincidental meeting of the Bethe Panel—and were highly interested in the significance of the new data.

After reviewing all available data, the Panel developed a list of 15 "Agreed Technical Conclusions" and outlined "Procedures for Estimating Identification Capability of (the) 170-Station Net." These conclusions essentially confirmed the interpretation of the Hardtack II data previously given here, and defined the parameters to be used for estimating the capabilities of a seismic network designed to represent the Experts' network of Control Posts. My staff began the analysis according to the Ad Hoc Panel's agreed procedures and conclusions.

On reflection, I suspect that the Panel's lengthy meeting was paced in good measure by this analysis. This was the era of paper and pencil hand-computations. Assessing the capability of a seismic network required the same laborious procedures as used during the Conference of Experts. We selected six earthquakes to represent the world's seismicity; and distances had to be calculated from each epicenter to all stations closer than about 30-40 degrees. First motion amplitudes had to be looked up for each distance and compared to the estimated noise amplitudes at each station for each epicenter, as well as for several explosion yields at each epicenter. Detection probabilities had to be calculated and so on. It required many man-days of work.

Key conclusions of the Ad Hoc Panel were:

a. The principal method recommended by the Geneva Conference of Experts for distinguishing earthquakes from explosions is of less utility than estimated prior to Hardtack II. That is, the determination of the direction of first motion is much more difficult than anticipated.

b. Additional information on the relationship between magnitude and equivalent yield in kilotons indicates that previous estimates of the number of earthquakes per year equivalent to a given yield in kilotons were low by a factor of about two.

c. As a consequence of conclusions a and b, and a revised estimate that a 3/1 signal-to-noise ratio is required to determine the direction of first motion, the number of earthquakes unidentified by the system recommended by the Geneva Conference of Experts is found to be as indicated in Table 7.1.

As a result of the conclusions in paragraph c, including Table 7.1, statements by the Geneva Conference of Experts concerning the detection and identification of earthquakes equivalent to 5 kt apply more nearly to earthquakes equivalent to about 20 kt.

The Panel recommended that additional explosions should be conducted to explore other methods of identification, to understand the coupling of explosions in other geological environments, and to investigate methods for concealing explosions. They also recommended that consideration be given to improving the detection and identification capability of the control posts recommended by the Conference of Experts by increasing the number of detectors in the arrays and by other means, and that such techniques should be tested as soon as possible.

"New Seismic Data" Reported in Negotiations

The Ad Hoc Panel's classified report was forwarded through the Department of Defense to the other Agencies involved in the test ban treaty negotiations, and to the President's Science Advisor, Dr James R. Killian, Jr. It was clear to him, and to the other members of the Committee of Principals, that the new data were highly relevant to the treaty negotiations in Geneva, and that the U.K. and U.S.S.R. should be advised of our results as soon as possible. Members of the U.S. Delegation in Geneva returned to Washington at about that time during a Christmas and New Year's recess. Over the holidays, a report to the Conference on the Discontinuance of nuclear Weapons Tests was drafted, outlining the implications of the new data for the negotiations. After extensive review within the government, the report, titled "New Seismic Data" was tabled 5 January 1959 as negotiations resumed in Geneva. The report contained tables and figures used by the Ad Hoc Panel, and was supported by photographic copies of 36 seismograms. The report states "stations were all equipped with Benioff short period vertical seismographs (described in the conclusions of the Geneva Conference of Experts)," asserting our understanding of the meaning of the ambiguously phrased conclusion previously discussed. It went on to provide our new understanding of the magnitude versus yield relationship, and its implication for numbers of earthquakes equivalent in size, our new findings on detectability of first motion, and so on.

A few hours after the report was tabled in Geneva, the American public was informed of the new developments through a White House press release. Key parts are reproduced below:

THE WHITE HOUSE

THE FOLLOWING STATEMENT ON THE DETECTION AND IDENTIFICATION OF UNDERGROUND NUCLEAR TESTS HAS BEEN PREPARED BY THE PRESIDENT'S SCIENCE ADVISORY COMMITTEE AND HAS RECEIVED THE CONCURRENCE OF THE DEPARTMENT OF STATE, THE DEPARTMENT OF DEFENSE, AND THE ATOMIC ENERGY COMMISSION. IT IS BASED ON CONCLUSIONS REACHED BY A PANEL OF SEIS-MOLOGISTS APPOINTED ON THE RECOMMENDATION OF THE CHAIRMAN OF THE PRESIDENT'S SCIENCE ADVISORY COMMITTEE.

Since the Geneva Conference of Experts last summer, United States seismologists on behalf of the Government have continued to study all available data on the problem of detecting and identifying underground explosions, including new data obtained from the underground tests conducted in Nevada this past October. These studies and new data indicate that it is more difficult to identify underground explosions than had previously been believed.

The Geneva Conference of Experts last summer concluded that: although it is not possible to identify an underground explosion by seismic means alone, it is possible to identify a large fraction of seismic events as natural earthquakes when the direction of first motion of the seismic signal is observed at several, appropriately located stations. This procedure reduces the number of seismic events which would be unidentified and could, therefore, be suspected of being underground tests. Analysis of all available seismic data on underground tests, including the data new since last summer, has shown that this method of distinguishing earthquakes from explosions is less effective than had been estimated by the Geneva Conference of Experts. These analyses and new data also indicate that the seismic signals produced by explosions are smaller than had been anticipated and that there are consequently about twice as many natural earthquakes equivalent to an underground explosion of a given yield as had been estimated by the Geneva Conference of Experts.

These two factors mean that there will be a substantial increase in the number of earthquakes that cannot be distinguished from underground nuclear explosions by seismic means alone...

This was followed on 16 January 1959 by a Department of Defense News Release, providing considerably more technical detail. It also named the Ad Hoc Panel members, including the four scientists previously mentioned who had attended only part of the meetings.

Rebuttal in *Pravda*

The Russian response followed swiftly. Under the headline "New Attempts by the United States of America to Create Difficulties in the Negotiations on Cessation of Nuclear Weapons Tests," An article by Yu. Riznichenko and F. Brekhovskikh (referred to here as R & B) was published in *Pravda* on 20 January 1959. They scathingly rejected the relevance of the data, arguing along the following lines:

> ...In this connection, the authors assert that this instrumentation corresponds to the recommendations of the Geneva Conference of Experts.
>
> This is not correct. The Conference of Experts recommended that the control posts should be equipped with an array of 10 vertical seismographs having a magnification of the order of 10^6 and having a maximum sensitivity for registering oscillations at a frequency of about 1 second with a sufficiently broad receiving band. In addition installation of several arrays of instruments with a broader band was recommended.
>
> Yet the American seismologists, as is evident from the above-mentioned paper and from the materials attached thereto, used an instrumentation with a considerably smaller magnification, with a narrow band, and which had a maximum oscillation sensitivity with a period of about 0.3 seconds. There was no such broad band instrumentation. It is clear to specialists that such apparatus does not correspond to the recommendations of the Conference

of Experts and has a lesser sensitivity, particularly with regard to registering the amplitude of the first motion of the P wave, which is an important factor in determining the period of the source of a seismic event.

There is no doubt that the application of the apparatus actually recommended by the Conference of Experts would have confirmed the conclusions of the experts concerning the magnitude of the seismic signal received from an explosion of a given yield.

There are several points to respond to here. First, the U.S. report did not "assert that this instrumentation corresponds to the recommendation of the Experts" in the broad sense that R & B chose to construe our statement. It merely laid out our position that the Benioff seismograph conformed to the Experts' requirements for *one* of the Experts' instruments, namely, the short period instrument. Second, the broad-band instruments they report as lacking in our program (true) are irrelevant to measuring either the magnitude or the direction of first motion—the basic measurements underlying our report—for explosions as small as *Logan* and *Blanca*. Third, the Benioff had *available* a gain of 10^6, but could not usefully be operated at such gains under noise conditions that existed at the sites during the experiment, nor could any other instrument. And fourth, if properly operated, arrays do not change the measured amplitude—either of the first motion or of the maximum in P used to calculate magnitude. R & B were making legalistic, but technically irrelevant, objections that we would hear again later in 1959.

R & B continued with a protest against our revision of the Rainier magnitude estimate and an assertion that we had ignored the data from the numerous permanent seismic stations.

It should be noted that the value of the seismic intensity from the *Rainier* explosion has been regularly decreased by the Americans from the 4.6 value published in the first scientific report of 1958 down to 4.3, which appeared during the Geneva Conference of Experts, and finally down to 4.1 which is presented by the authors of the paper under consideration. Actually it is on this decease of the value of seismic intensity from underground nuclear explosions that the authors of the paper base their assertion that there are great difficulties in detecting explosions on the basis of seismic data and that there are many natural earthquakes that allegedly cannot be distinguished from explosions. The authors of the paper have not examined the data registered on the *Blanca* explosion at other seismic stations in the United States, which total over 90, as well as stations in Sweden and in other countries located at distances of several thousand kilometers from the epicenter.

Shortly after the *Rainier* shot a preliminary report of magnitude 4.6 had been made in 1957, based on scanty, unverified station reports. It was never used in any official way by the U.S. in test detection discussions with the USSR. The first authoritative report was by Les Bailey and myself early in 1958; we had estimated *Rainier*'s magnitude at 4 1/4, and this was the value reported to the Experts. The pattern of "regularly decreased" magnitude for *Rainier* did not exist in the official U.S. estimates. Furthermore, the downward revision in magnitude of this one shot had little to do with our revision of the magnitude vs. yield relationship, which in turn, had led to our increased estimate of the numbers of earthquakes equivalent to explosions of given yield. Rather it was data from *Tamalpais* and other low yield shots that caused the revision (see Figure 7.4). And the assertion by R & B that we had not examined data from other seismic stations was simply false -- not only had we examined the data, but the USC&GS had conducted an independent analysis of it.

R & B go on to state:

> The authors also fail to take into account the important fact that the *Blanca* explosion was accompanied by rupture of the earth's crust and by a strong venting of material and, consequently, the energy of its seismic effect must be smaller than in the case of an underground explosion of the same yield. Thus, all these far-reaching conclusions are actually based on the data from only one new explosion, i.e., the *Logan* explosion. Likewise, for reasons unknown, the authors failed to consider the readings on the *Rainier* explosion taken at many other stations.
>
> Far from "a strong venting of material" that might have vented energy into the air, thereby reducing seismic energy, the rupture was actually a collapse into the underground cavity created by *Blanca*. Since it occurred about 15 seconds after the explosion, the seismic waves were unaffected because they were already 50-100 km from the explosion site at the time of the collapse. Readings from the *Rainier* explosion at other stations not equipped with the Wood-Anderson seismic graphs specified for magnitude estimates were, in fact, used for other kinds of analyses.

And thus the lines for future Soviet arguments were laid out.

EIGHT

The Berkner Panel

The report of the Ad Hoc Panel of Seismologists moved quickly through the U.S. Government in late December. Their key conclusion was that "statements by the Geneva Conference of Experts concerning the detection and identification of earthquakes equivalent to 5 kt apply more nearly to earthquakes equivalent to about 20 kt." This conclusion, contradicting the technical basis underlying on-going negotiations in Geneva, caused concern at the Department of State, which was responsible for overseeing those negotiations. Perhaps the Ad Hoc Panel's recommendations on research to improve the Experts' system also caught the eye of State Department officials. However it happened, the result was that the President's Special Assistant for Science and Technology, Dr. James Killian, appointed a "Panel on Seismic Improvement " on 28 December 1958 at the request of the Secretary of State, at that time Christian Herter.

The new panel was to be chaired by Dr. Lloyd V. Berkner, President, Associated Universities Inc., and a member of the President's Science Advisory Committee. The Panel commonly referred to itself as the "PSI," but subsequently became better known as the "Berkner Panel." Berkner was an impressive figure with broad experience in science. He had served in several agencies of the U.S. Government, the Carnegie Institution in Washington, and the Joint Research and Development Board. He had been a member of Admiral Byrd's Antarctic expedition of 1928-30, and a key organizer of the International Geophysical Year. At the time of the panel meetings, Lloyd Berkner was vice-president of the American Geophysical Union. He specialized in physics of the upper atmosphere. Other members were:

Hugo Benioff, California Institute of Technology
Hans A Bethe, Cornell University
W. Maurice Ewing, Columbia University
John Gerrard, Texas Instruments, Inc.

David T. Griggs, University of California at Los Angeles
Jack H. Hamilton, The Geotechnical Corporation
Julius P. Molnar, Sandia Corporation
Walter H. Munk, Scripps Institute of Oceanography
Jack E. Oliver, Columbia University
Frank Press, California Institute of Technology
Carl F. Romney, Department of Defense
Kenneth Street, Jr., University of California
John W. Tukey, Princeton University and Bell Telephone Laboratories.

Members of the PSI thus represented expertise in solid earth geophysics, oceanography, nuclear physics and chemistry, engineering, mathematics and digital data processing (an emerging new field at the time).
The Panel's charter was:

> The Panel should determine whether it would be reasonably feasible within the present state of seismic technology to improve the capabilities of the system recommended by the Geneva Conference of Experts to detect and to identify seismic events as either earthquakes or explosions without increasing the number of manned control posts in the system. The Panel's investigation should include, but need not be limited to, the following: (a) improvements or augmentation of equipment at control posts in the agreed Geneva system; (b) augmentation of the system with a more closely spaced grid of small, completely automatic seismic detectors; and (c) utilization of criteria other than the first motion of the P wave to identify events as earthquakes (or as explosions).
>
> The Panel should also recommend a research and test program to evaluate any specific proposals advanced to improve the system, as well as to advance the state of the art in this field. The Panel should indicate the extent to which nuclear tests would be required in this test program.

Panel meetings

Anticipating that the "New Seismic data" would evoke a negative reaction by the Soviets when tabled in Geneva, the PSI was urged to proceed swiftly. Accordingly, the first meeting convened on 6 January 1959—the day after the new data were presented at the negotiations. The Panel met in Room 272 1/2 of the Executive Office Building of the White House, where the members were briefed on seismological aspects of the Experts' report, and I presented the Hardtack II data, along with the Ad Hoc Panel's conclusions on their implications. After reviewing the assumptions and methodology of the Ad Hoc Panel's study, the Berkner Panel endorsed its main conclusions, and slightly revised the estimated numbers of unidentified earthquakes (but well within the range of uncertainty given by the Ad Hoc Panel).

We then began to suggest and discuss potential ways to improve the Experts' system, and in the process, to identify deficiencies in data or understanding that impacted our ability to improve the system. A wide spectrum of ideas, ranging from new types of sensors to studies of earthquake physics, and from detonating large underground explosions to the application of advanced data processing methods, were suggested, discussed, and eventually reduced to a list. All were promising, but none were immediately applicable to the Experts' system without further investigation or development.

The Panel adjourned on 7 January 1959 after preparing an interim report to Dr. Killian. The report pointed out that the method of identification of seismic events used by the Experts' system "places principal reliance on a single phenomenon, the direction of displacement of the first motion of the P-wave." It went on to mention that the PSI had considered a variety of other phenomena and methods that might increase the capability of the system without adding control posts, and it listed more than a dozen. These included methods requiring fundamental research, as well as more straightforward engineering enhancements of systems for detecting and analyzing seismic data. The Panel noted that it believed that seismic research had not been supported as strongly as many other areas of science, and that vigorous research "is certain" to produce many improvements. It urged that a scaled down experimental network of seismic stations similar to the proposed Geneva system be established in the United States without delay, and that the existing worldwide network of seismic stations be upgraded "within the next year, even if it must be done unilaterally by the U.S." In response to the charter's question on the extent to which nuclear explosions were needed, the Panel's answer was clear: a number of nuclear explosions would be required, several as a matter of high priority. This conclusion added momentum to a proposal, tabled at the formal treaty negotiations in Geneva on 30 January, that is, that nuclear detonations be permitted for peaceful purposes under appropriate safeguards. The main force behind the proposal was the U.S. "Plowshare Program," an ambitious plan to employ nuclear explosions to excavate major new canals, stimulate oil and gas fields by fracturing rock, creating deep underground cavities for storage of petroleum or toxic waste, and so on. The proposal was rejected a month later by the Soviets.

Writing assignments on the Panel's listed topics were accepted by several members (for the next meeting) to develop these ideas more fully, and to outline the necessary research. For example, John Gerrard, working with Jack Hamilton, was to describe an approach to developing unattended seismic stations; Jack Oliver was to describe potential use of seismometers placed on the ocean bottom, and I was to outline a program for investigating short period shear waves. Kenneth Street, Deputy Director of the Lawrence Livermore Laboratory, undertook the task of explaining the need for underground chemical and nuclear explosions and to outline a program of tests designed to explore the associated seismological problems. A program of particular interest to seismologists generally was subsequently developed by David Griggs and Frank Press: they recommended that "the 100-200 best seismic stations in the world be equipped with modern instruments as soon as possible." Imagine the benefit to our Hardtack II seismic program if the existing permanent stations in the U.S. had been equipped with modern, well-calibrated seismographs giving accurate amplitude measurements!

Additional full Panel meetings were held in Dallas, Texas, and New York City, concluding in Washington again. The Dallas meeting was hosted on February 9-10 by The Geotechnical Corporation at their Haggar Drive offices, and by Texas Instruments, Inc. at their Lemon Avenue offices. At Texas Instruments, many of us saw our first large-scale digital computer. It literally filled a large laboratory with equipment racks supporting hundreds of electronic chassis containing thousands of vacuum tubes. It was impossible to envision future home computers evolving from that gigantic model! The Panel reviewed initial drafts of research programs outlined by members, and made preliminary cost estimates for a two year program costing $26.5 million the first year and $29.7 million the second. Of these amounts, $12 million each year was proposed for large explosions. A third meeting with similar agenda took place on March 5 and 6 at the offices of Associated Universities, Inc., 10 Columbus Circle, New York City.

Decoupling

Even as the Panel was devising improvements to the Control System to compensate for deficiencies revealed by Hardtack II, the Panel members were about to be faced with a still greater challenge. Dr. Edward Teller and his associates at the Radiation Laboratory in Livermore, California continued to believe that clandestine testing methods had been given too little consideration. Teller was a frequent visitor to AFOAT-1 at the time. He was a man of firm belief that the Soviet Union would cheat on any nuclear test ban if they could find a way to do so. I remember several discussions in Doyle Northrup's office listening to him expound on his views, and watching as he stalked back and forth at Doyle's blackboard, punctuating his remarks with a diagram, or perhaps an equation. He was determined to understand how the Soviets might cheat. His associates back in Liver-more also began a study of potential evasive testing methods, including re-examination of various suggestions that had previously been made. In the process, Teller's colleague, Dr. Albert Latter of the Rand Corporation, recognized a flaw in his prior thinking about the influence of the size of the chamber in which an explosion might take place.

In their recently completed paper that attempted to explain the ob-served first power relationship between amplitude and yield of the Hard-tack II shots, Latter, Martinelli and Teller (1959) had also shown that for any given yield, amplitudes of low frequency seismic waves should be independent of the size of the cavity in which the explosion took place. This theory, however, assumed that the cavity walls moved elastically. Substantially the same elastic theory had been published in the United States by J.A. Sharpe (1942) who, in turn, cited an even earlier publica-tion by Kawasumi and Yasiyama. The theory would apply if the cavity were large enough that the pressure created by the explosion did not produce motions exceeding the elastic limits of the surrounding rock. In practice, the few underground nuclear explosions to date had been fired in small chambers and pressures had greatly exceeded the elastic limits of the rock.

Latter and his associates, R.E. LeLevier, E.A Martinelli, and W.G. Mc-Millan (1959) applied the previously mentioned theory to calculate the P-wave amplitude expected from *Rainier,* had it been fired under elastic con-ditions. Measurements of the actual amplitudes had been made, however, and when Latter, *et. al.,* compared them with the theoretical amplitudes, the measured amplitudes were found to be about 50 times larger. Since there was (and is) no reason to doubt the elastic theory, the conclusion was clear and simple: had the *Rainier* shot chamber been large enough to keep the pressure below the elastic limit of the rock, its signals would have been 50 times smaller. And thus was born the idea of "big hole decoupling" as a way to conceal an explosion by making its seismic signals undetectable.

The authors calculated that an even greater signal reduction would occur if the cavity were to be constructed in a more rigid material; they predicted a signal reduction, relative to *Rainier,* of about 300 for a cavity in salt. They suggested as a design criterion that the cavity be made large enough that the equilibrium pressure following the explosion would be about one-half the pressure due to the weight of the rock above the cav-ity. This would require a diameter of about 180 feet for an explosion as large as *Rainier.* While this seems large, much larger cavities had been constructed in salt domes (by dissolving the salt in water) for storage of petroleum products. Construction was conceptually simple: water was pumped down a drilled hole, circulated to dissolve the salt, and then the brine was pumped back up through a separate pipe and disposed of. It seemed that cavity decoupling might be an effective method of circumventing detection by the Experts' system.

In due course the concept was presented to the Berkner panel. One of the Panel's members, Hans Bethe, independently reviewed the theory and the data, and reported his conclusion that the amount of decoupling might be as much as 400. He also reported that the cavity need not be spherical. The Panel viewed decoupling with caution, however, recognizing the need for experimental verification (as had Latter and associates).

Final Reports of PSI

The SECRET level final report of the Panel on Seismic Improvement was submitted to the President in two parts on 16 and 24 March 1959. It contained both good news and bad news. The good news in the findings addressed the first paragraph of the Panel's charter. In summary, the Panel concluded: "Equipment and techniques can be specified today that would give the Experts system the same capability as originally estimated for events ≥ 5 kt." The report also included a table, estimating the annual number of unidentified seismic events under various assumptions about the equipment at control posts and its effectiveness, as shown in Table 8.1.

The table applied to earthquakes equivalent in size to explosions detonated under *Rainier* coupling conditions. Numbers of earthquakes were uncertain by at least a factor of two; that is, actual numbers might be twice as large or half as large as estimated.

The bad news was in the second classified report giving our findings on concealment. The Panel reported that we had examined the possibility of concealment by using a suitably designed very large shot chamber. Its conclusion was that "techniques existed that could reduce the seismic signal by a factor of ten or more," and we noted that "The seismic signal from one Hardtack II test (*Evans*) was x 10 less than that from another (*Tamalpais*) of approximately the same size, although no attempt was made to reduce the signal." Moreover, "theory shows that it is possible in principle to reduce the seismic signal by a much greater factor." However, the panel cautioned that the theory must be tested.

The improvements indicated in the preceding table (reduced numbers of unidentified events in the third estimate) resulted primarily from increasing the number of sensors in the array from 10 to 100, and an assumption that this would improve the detectability of P-waves by about a factor of 8. This assumption required a significant act of faith in hypotheses about the nature of seismic noise. It was known that in the signal pass band, taken to be about 0.3 to 5 Hz, there would be significant correlation of noise among the 100 sensors deployed over an area 3.5 km in diameter, at least at the lower frequencies. And it was hypothesized that this correlation could be exploited to discriminate between signal and noise.

"Velocity filtering"—enhancing or rejecting seismic waves according to their speed and direction of arrival at the array—was one of the tools proposed. This, and other analytical tools considered, would require the design and development of special purpose digital computers. And while the required computing technology existed—at least in principle—working models would have to be built and tested at several locations just to determine if noise and signals actually had the assumed properties. In short, the proposal for installing 100 sensor arrays at the Experts' Control Posts was based on little more than the enthusiasm of advocates of array technology. We now know that the results would have been highly disappointing (see Appendix F).

The Proposed Research Program

On 31 March 1959 the Panel submitted its detailed unclassified report titled "The Need for Fundamental Research in Seismology." The main body of the report consisted of a collection of papers authored by individual panel members, primarily outlining research projects. Jack Oliver, Frank Press and I wrote a "Summary Report" as an introduction to the complete report, synthesizing the individual projects and providing a context for the various research proposals. The Summary Report was published in a scientific journal in September with a short preface by Frank Press (1959).

The summary contained a sales pitch for research. It contained a two year, multimillion dollar cost proposal. Major elements of the proposed research and development programs are shown in Table 8.2.

Our summary pointed out the modest level of funding for classical seismology in the past—perhaps several hundred thousand dollars annually in the U.S. And it contrasted this with estimates of much larger funding in the Soviet Union (a ploy that I knew federal officials were tired of hearing from scientists). But the summary also explained the problems associated with applying seismology to nuclear test detection.

Concerning seismic discrimination, the report states:

> We hope to learn, from seismic evidence, how to distinguish between natural earthquakes and explosions, at least in the great majority of cases. If we are to do this, it is essential that we improve our knowledge of the phenomena associated with the various types of sources of seismic waves. We need to know more about the mechanism of seismic wave generation when the source is a nuclear or chemical explosion, but it is equally important,

perhaps more so, to understand the mechanisms of all kinds of natural earthquakes.

It went on to point out the paucity of critical data:

Our knowledge of seismic wave generation by large explosions, particularly nuclear explosions, is very limited. Only three completely contained underground explosions with yields greater than one kiloton have been fired, all under very similar environmental conditions. The parameters that can significantly affect the magnitude and type of seismic effects produced by nuclear explosions should be experimentally explored so that theories can be developed which will permit reasonable confidence in our understanding of these parameters and deductions about them. The following parameters require study:
1) dependence of spectrum of body waves and surface waves on yield of explosion;
2) dependence of seismic wave excitation on the medium surrounding the shot;
3) effect of depth of burial;
4) effect of local environment such as shot-cavity size and shape, etc.;
5) effect of local geology and topography.

And it recommended:

The following experimental nuclear shots should be carried out as soon as feasible: (a) a 5 kt shot in granite for information on the effect of shooting in another medium, (b) a shot in an environmental situation designed to decouple explosion-energy from seismic energy, (c) two 5 kt shots near the *Rainier* site but at appreciably greater depths. Theoretical studies which suggest the possibility of concealment by reduced coupling should also be experimentally tested as soon as feasible.

The summary pointed out the problem of "inadequate, non-uniform instrumentation" at existing seismic stations, and endorsed re-equipping 100-200 of the world's stations with modern instruments. A vigorous program to improve seismic detection methods was proposed, including placing seismometers below the earth's surface in bore-holes—a technique soon to produce highly significant results in noise reduction, even if not at the great depths initially proposed. The panel had placed great emphasis on the immediate need to construct a complete experimental station incorporating all features of the seismic stations recommended by the General Conference of Experts! And the summary proposed the then-novel idea of establishing a computer center for seismology! The Department of State subsequently published the full report, "The Need for Fundamental Research in Seismology."[22]

On 12 June 1959 Ambassador Wadsworth introduced both a declassified version of the findings of the Panel and the research document to the nuclear test ban treaty negotiations in Geneva. He proposed a joint study of the findings, which was rejected by the Soviet Ambassador Tsarapkin on the grounds that such a study would delay the negotiations and that, anyway, the detection system could be perfected after it went into operation.

Aside from the brief hint of possible "reduced coupling" cited above, and mentions of "shot-cavities size and shape," the concept of decoupling by means of a large cavity remained classified SECRET until October of 1959. This matter was, however, discussed with the British during a classified meeting in London on the 10th and 11th of August. The seriousness of the meeting was made evident to the British by the group's leader, the President's science advisor, James Killian. I felt fortunate to be seated next to him and to share views of the meeting on the return trip. It was also my first round trip across the Atlantic in a jet aircraft.

Instituting the Proposed Research Program

The Panel's report was circulated and discussed by agencies within the U.S. Government concerned with test ban treaty negotiations. As a result, a decision was made to implement the proposed research and development program. On 27 April 1959 the President's Science Advisor, Dr. Killian, and senior representatives of the Chairman of the Atomic Energy Commission, John McCone, and the Deputy Secretary of Defense, Donald Quarles, agreed that the Department of Defense would serve as the lead agency, with support provided by the AEC.

As a consequence, on 7 May 1959 General Herbert Loper, Assistant to the Secretary of Defense for Atomic Energy, assigned responsibility for direction of the program to AFOAT-1. In response, we in AFOAT-1 advised the U.S. seismological community of this new program, and invited proposals to participate. With proposals in hand from numerous universities, and industrial and government laboratories, we prepared a program plan listing specific projects for implementing the Berkner Panel's recommendations, and on 2 July 1959 this plan was sent to the office of the Secretary of Defense (General Loper). Three weeks later we followed this with a request for slightly more than one million dollars of Emergency Funds to initiate the program. General Loper's office, however, had no funds for the purpose, and the plan and funding requests were forwarded to the Director of Defense Research and Engineering, Dr. Herbert York.

York promptly appointed and convened an advisory panel chaired by Dr. Frank Press, and largely drawn from former members of the Berkner Panel. This "Ad Hoc Group on Seismology" reviewed and approved our program plan on 17 August. However, Dr. York recommended that the program be assigned to the Advanced Research Projects Agency (ARPA), which the Secretary of Defense authorized on 2 September 1959, stating that he expected ARPA to assign significant technical portions of the program to AFTAC.[23] (AFOAT-1 had ceased to exist on 7 July 1959, and the organization had been renamed the "Air Force Technical Applications Center" (AFTAC) the same day.)

We promptly requested Emergency Funds from ARPA to initiate the program and on 2 October ARPA Order 104-60 provided $460,000 to AFTAC to do preliminary work on a few of the projects. It was a weak beginning compared to what we had hoped and the Berkner Panel had recommended, but at last work could commence on the Vela[24] Program, as ARPA had named the seismic research program.

NINE

Technical Working Group II

Treaty Negotiations and The New Seismic Data

In Geneva, diplomatic negotiations within the "Conference on the Discontinuation of Nuclear Weapons Tests" (hereinafter "Conference" or "political conference") continued, after the U.S. tabled the "New Seismic Data" in January, 1959. But the Soviets, through Ambassador Tsarapkin, continued to reject the new data, insisting that the report of the Experts was the sole technical basis for the negotiations. U.S. Ambassador James J. Wadsworth was equally insistent that the new data must be considered, and he proposed a technical meeting to assess the data and advise the Conference on its implications. Ambassador Tsarapkin flatly refused, dismissing the U.S. document as "preliminary and hastily prepared" and not requiring study. He also charged that such a meeting would simply delay the negotiations. He similarly rejected Ambassador Wadsworth's proposal for a technical meeting to discuss methods for detecting high altitude explosions (which the Experts had only considered briefly), stating that both seismic and high altitude matters could be dealt with by a "Control Commission" to be established under the treaty.

However, a major consideration for the Soviets must have been the new seismic data's implications for on-site-inspections. At the opening of the negotiations the Soviets had tabled a short draft treaty that was silent on inspection, except perhaps implicitly by the mention of instituting "machinery for control...in accordance with the recommendations of the Geneva Conference of Experts." Their position was that the three nuclear powers should agree as quickly as possible to a permanent ban on all nuclear tests, leaving matters like the control system to be settled later by the Control Commission.

The Western nations, on the other hand, believed that agreement on an effective system of controls was an essential precondition to accepting a permanent ban on testing. The Western position was that specific features of the system of control should be spelled out as part of the treaty. Consequently, on 16 December 1958 Ambassador Wadsworth had proposed a draft Annex to the treaty dealing with the installation, operation, and improvement of the control system, including on-site inspection. It called for inspection of all unidentified seismic events with an estimated magnitude equivalent to an explosion of 5 kt or greater, and of 20% of those smaller than a 5 kt equivalent. Any event was to be inspected if the seismic data suggested the event might "have an unusually high probability of being of nuclear origin." This latter point apparently left the door open for application of the long-sought explosion "signature," in case it were to be discovered in the future.

When coupled with the proposed U.S. formula for selecting events for inspection, the new data suggested there might be more than 1,000 unidentified events worldwide larger than a 5 kt equivalent. Even after achieving the capabilities projected by the Berkner Panel, the system would still leave more than 300[25] unidentified events annually. Only a fraction of these would be within the territory of the Soviet Union but, nevertheless, the increased number could require a proportionately larger opening up of Soviet territory than they had previously contemplated. Ambassador Tsarapkin was direct about this, charging that the new data had been introduced only to justify more inspection parties, and that "inspection then, to put it bluntly, becomes intelligence work."

Given the Soviets' near-paranoia about foreign inspection of any part of their territory, their responses were almost predictable. Not only did they refuse to accept the new data, they soon proposed that a number of important actions to be taken under terms of the treaty should require "unanimity" among the principal parties. Unanimity was the sugar-coated term used by the Soviets at the time to obscure its corollary: any party to the treaty would be able to *veto* any activity covered by the term. The Soviet proposal covered some actions that appropriately called for unanimity, e.g., revisions of the treaty itself, but they also proposed that it be applied to sending out inspection parties for on-site investigations as well as to making decisions on the results of the inspection. Obviously, in the Western view, the Soviets could not be given veto power over functions that were at the very core of the system of control.

A Breach of AFOAT-1's Security

On 20 January 1959 Doyle Northrup, at that time still a scientific advisor to Ambassador Wadsworth in Geneva, was awarded the President's Award for Distinguished Federal Civilian Service. In a ceremony at the White House his wife, Sybil Northrup, accepted the award from President Eisenhower, the highest award given to career civilian employees of the

U.S. Government. Northrup was one of only five employees selected to receive the award that year.

The citation accompanying the award was explicit and revealing. The award was for "his immense contribution to the security of the United States" in developing "our system of nuclear detection and surveillance." The secret mission of AFOAT-1 was clearly spelled out in the announcement the following day by the *New York Times* under a headline, "Lost: Cloak and Dagger." In Geneva, M.A. Sadovsky and others of the Soviet delegation promptly congratulated Northrup for his high award and for his achievements in developing a nuclear detection system, even though it chiefly targeted their own country.

Inspection Quotas and a Phased Treaty

During a visit to the U.S.S.R. in late February of 1959, British Prime Minister Harold Macmillan, picking up on a Soviet idea suggested by Fedorov, proposed that a quota of on-site inspections be negotiated as a possible means of breaking the deadlock. He mentioned numbers ranging from 3-5 to up to 20 annually. Soviet Premier Nikita Khrushchev neither accepted nor rejected the idea, but reacted favorably. The quota was discussed at a U.S./U.K. summit meeting in mid-March, but the idea was not well received by the U.S., where it was seen to be inadequate to resolve the expected number of unidentified events in the U.S.S.R.

On 13 April President Eisenhower made a new proposal to Premier Khrushchev. Noting that Soviet proposals to date did not provide for effective control, he concluded that there was no basis for agreement by the West on a comprehensive test ban treaty. He proposed an alternative:

> If indeed the Soviet Union insists on the veto on the fact finding activities of the control system with regard to possible underground detonations, I believe that there is a way in which we can hold fast to the progress already made in those negotiations and no longer delay in putting into effect the initial agreements which are within our grasp. Could we not, Mr. Chairman, put the agreement into effect in phases beginning with a prohibition of nuclear weapons tests in the atmosphere?

President Eisenhower pointed out that the necessary control system for atmospheric tests up to an altitude of 50 km could be quite simple, and would not require the on-site inspection that created "the major stumbling block in the negotiations so far." Such a treaty could be extended to underground tests, and to tests above an altitude of 50 km, after the technical issues associated with monitoring tests in these environments were resolved. Presumably, resolution would be reached through major research programs proposed by the Berkner and Panofsky Panels (both still under review within the government but soon to be approved).

Premier Khrushchev rejected the proposal for a phased treaty and countered that compromise might be possible on the basis of a fixed quota of inspections, citing his earlier discussions with Prime Minister Macmillan. The quota would apply to events that exhibited "phenomena that might be suspected of being nuclear explosions." Inspection would have to be based on "objective readings of instruments" of the control system. The problems associated with the proposed veto and high altitude tests remained unsettled, however.

Technical Working Group I

Progress toward resolving the technical issues came in May, when Premier Khrushchev agreed to a technical meeting on high altitude tests. The meetings, to become known as "Technical Working Group I" (TWG-I) took place in Geneva under the aegis of the Conference in June and July of 1959. The U.S. group was chaired by Dr. Wolfgang Panofsky, an eminent physicist from Stanford University. The Working Group reached agreement on a set of ground-based and satellite-based sensors to supplement the Experts' control system, and presented their recommendations to the Conference in July 1959. The meetings also set the precedent for a second technical working group to consider the new seismic data.

Objective Criteria for On-Site Inspection

On 12 June Ambassador Wadsworth presented the findings of the Berkner Panel to the Conference, and proposed a joint study of them. He pointed out that implementing the improvements to the Experts' System, as the Panel proposed, would increase the number of events identified by the system, thus reducing the number eligible for inspection. After an initial suggestion that experts might consider a study of identification criteria only, Ambassador Tsarapkin's final response was that the Experts' Report of 1958 provided figures on unidentified events and the Soviet Union would stick to them. Hence no new meeting of experts was needed. He later attacked the Berkner Panel's recommended unmanned seismic stations as potential vehicles for foreign agents inside the Soviet Union.

The Soviet Union continued to reject the new seismic data, but in early November, Ambassador Tsarapkin proposed the formation of a technical working group to formulate "objective criteria" for the dispatch of an on-site inspection team. The U.S. responded that both the New Seismic

Data and the recommendations of the Berkner Panel had significant implications for the formulation of such criteria, and thus should be studied. Eventually, on 24 November, the Soviet Union formally agreed to such a study in connection with their study of "objective criteria."

Technical Working Group II Begins

Technical Working Group II (TWG-II) convened on 25 November in Geneva, under terms of reference providing:

> The Technical Working Group of Experts shall consider the question of objective instrumental readings in connection with the selection of an event which cannot be identified by the international control organ and which could be suspected of being a nuclear explosion, in order to determine a basis for initiating on-site inspections. As part of their work, the experts, proceeding from the discussions and the conclusions of the Conference of Experts, shall consider all data and studies relevant to the detection and identification of seismic events and shall consider possible improvements of the techniques and instrumentation.

Dr. James B. Fisk chaired the U.S. delegation, with Dr. Wolfgang Panofsky, who had been the U.S. Chairman of TWG-1, serving as Deputy Chairman (Figure 9.1). Most of the 14-man U.S. team had also been representatives at the Conference of Experts; apparently their selection was intended to send a message to the Soviets that the U.S. regarded the new working group to be a continuation of the Experts' work. The U.S. seismologists were Drs. Norman Haskell, Air force Ambridge Research Center; Jack Oliver, Columbia University; Frank Press, Caltech; and Carl Romney.[26]

The Soviet team was chaired once again by Dr. Yevgeny Fedorov, and included seismologists V. I. Keilis-Borok, I.P. Pasechnik and Yuri V. Riznichenko. M.A. Sadovsky and several other familiar faces from the Experts' Conference were also present (Figure 9.2). The British team was chaired by Sir William Penney; it included no seismologists, but young physicist J.K. Wright, who was beginning to work in the field, attended (Figure 9.3). He would later make a number of excellent scientific contributions to seismology.

After some wrangling over an agenda, and introductory remarks by the heads of the three delegations, the working group turned to consideration of the new seismic data. This was preceded, however, by Dr. Sadovsky's statement that, based on the data given to them months earlier, the Russian delegation saw no reason to revise the Experts' Report or its conclusions—a theme repeated in several forms many times throughout the meetings.

My presentation of the new data and its implications—essentially along the lines agreed to by the Ad Hoc Panel of Seismologists and the Berkner Panel—was met promptly by questions and by critical comments from Fedorov. He impatiently asked if all of the seismic data had been presented, obviously anxious to reject them and move on to other topics. When we pressed him for any data the Soviets might have on Hardtack II, Fedorov hinted that they might have data, but offered none.

During the second meeting the next day Jack Oliver and Frank Press described the Love waves and shear waves recorded from *Logan* and *Blanca*, pointing out that these were within the amplitude range of earthquakes having similar P-wave magnitude. Such waves had been claimed to be positive indicators of earthquakes during the Experts' Conference; accordingly, these data, by themselves, were sufficient to call for a review by the Conference. We presented 250 additional seismograms recording Hardtack II shots to the Soviets, once again requested any relevant seismic data they might have, and urged them to examine all data jointly with us to reach agreed conclusions. They presented no data and refused the joint study. Instead, Fedorov accused us of selectively providing data intended to cast doubt on the Experts' conclusions, and he implied that we were withholding other data more favorable to their case.

The third meeting brought the moment Fedorov seemed to be waiting for. Reading from a written statement that signaled the character and tone of his comments throughout the remainder of the conference, he sarcastically denied the significance of the new data and categorically rejected our conclusions based on them. The technical elements underlying Fedorov's and subsequent Soviet criticisms were mostly those presaged by the Riznichenko/Brekhovskikh letter to Pravda, previously discussed. The weight of their arguments was not primarily technical, however, but rather legalistic and formalistic, designed to disqualify our data as inadequate to support our conclusions. U.S. scientists rapidly reached the conclusion that the Soviet scientists were under strict political guidance not to agree to any point that might undercut the Soviet stance in the diplomatic Conference, namely, that the Conference of Experts Report was both correct and adequate for negotiating a treaty.

Fedorov's statement was also replete with distortions and misrepresentations. He said, for example:

> In the working papers of 5 January it is stated—and Mr. Romney also said in his presentation—that the Geneva Conference of Experts based its conclusions regarding underground explosions

on data from only one experiment: the Rainier explosion... the 1958 discussion was based on the experience of worldwide seismology, both as regards natural earthquakes and as regards...hundreds of explosions.

Although we had pointed out that there had been only one previous underground nuclear explosion, neither the working paper nor I had made the obviously absurd statement that Expert's conclusions were based only on *Rainier*. Nor had either dismissed or denied the valuable information obtained from earthquakes and chemical explosions; but both were irrelevant to the subjects of our new analysis: the magnitudes, amplitudes of first motion, and the types of seismic waves generated by deep underground nuclear explosions.

Calibration of Permanent Stations

An early criticism by the Soviets was that we had not used data from many permanent U.S. seismic stations. I responded that we had used the data for other purposes, but could not use the data for measuring magnitudes because the stations were not calibrated. We argued that, while additional calibrated data might be useful, the calibrated data we had given to them were sufficient to prove the conclusions we had derived from them. Nevertheless, Fedorov persisted in assertions implying that the existence of other uncalibrated data demonstrated poor management of our experiment, and that in some unexplained way, this somehow clouded the value of our primary data.

However, as organizers and implementers of the long range measurements experiment, we in AFOAT-1 had neither the authority to pressure the various universities to calibrate their seismometers, nor the manpower to help them—even if security had permitted us to do either (nuclear testing plans were secret). Beyond these factors, we knew that amplitude measurements—our primary objective, and the feature that made our experiment unique—were sensitive to the specifics of the seismograph's response curve, and thus measurements from other types of instruments with different response characteristics would only corrupt the data. (It would be several years before U.S. stations would have magnetic tape recordings, and we would have computers able to correct the recordings to common response characteristics.) But the Soviet focus on data we had not used for estimating magnitudes was, to them, an excuse for refusing to consider the valid data we had provided to them.[27]

Seismographs and Arrays

Similar legalistic objections to the data rapidly developed as the Soviet scientists asserted that we had not deployed the seismographs recommended by the Conference of Experts for control posts, and therefore our data could not be used to estimate the capabilities of the Experts' system. Principal criticisms were that we had not installed seismometer arrays at each station, and that our Benioff response characteristics were wrong. The Soviets ignored our reminders that the Conference of Experts also had no data from fully equipped control posts, yet somehow had been able to reach important conclusions based on data from the same type of instruments that we had used for Hardtack II, as well as from less capable instruments.

Dr. Fedorov was particularly harsh and persistent about the absence of arrays. He asserted that this absence invalidated our conclusion that our measured amplitudes of the first motion of the P-waves were smaller than we had estimated at the time of the Experts' Conference. My response that arrays would only improve the signal-to-noise ratio, but would not increase the measured amplitude of first motion fell on seemingly deaf ears. Subsequently, we pointed out that the *Blanca* signal-to-noise ratio on single sensors was almost exactly that expected for *Logan*, had *Logan* been recorded on 10-element arrays! Thus we could project with high confidence the amplitude of first motion generated by 5-kt shots and recorded by arrays. But Fedorov was obviously not interested in logic.

Fedorov continued his attack in abusive and dismissive tones. He was much my senior, and I was at a loss as to how I should cope with what seemed to me to be deliberate and blatant distortions of the facts. How in the world could a competent scientist face his own colleagues after making what even they would recognize as an absurd argument to reject valid measurements? True, we might have been able to obtain more measurements had we been able to deploy arrays, but that did not change or invalidate the measurements we had been able to make. Noting my growing frustration, Sir William Penney, leader of the British delegation and equal to Fedorov in experience and stature, calmly took him to task, pointing out that arrays improve the signal-to-noise ratio, but are not needed if the signal is already larger than the noise—as was true for enough of the Hardtack II stations to allow us to measure first motion at various distances. Fedorov dropped the criticism for the time during the meetings (but only to repeat it in a Soviet Annex to the final report of TWG-II).

As discussed earlier, the U.S. Experts had advocated short-period seismographs having the approximate response characteristics of the Benioff instruments employed in the AEDS. At exceptionally quiet sites, these had magnifications of up to 1,000,000 at 0.2 and 1.0 seconds with a peak in between. In the English language version of the Experts' Report, the words "the seismographs should have a maximum magnification of the order of 10^6 at a frequency of 1 c.p.s. and a receiving band adequate to reproduce the characteristic band of the seismic signal," although ambiguous in details, we had interpreted as describing, or at least including, a Benioff. No such interpretation was possible of the Russian language version—or so we were told. The peak of the response curve must be at 1 c.p.s., similar to the response of the Soviet SVK-M seismograph. In fact, Pasechnik strongly implied that the SVK-M fit the Experts' specifications. When challenged on this point by Dr. Fisk, Dr. Fedorov interjected that the SVK-M was not the instrument specified by the Experts, but that perhaps it conformed more nearly than did the Benioff. On several occasions we pointed out that an instrument with a flat peak at one second would necessarily have far too great sensitivity at a 5-6 second period in the main microseismic band; it could not possibly be operated at a gain of $10.^6$ Pasechnik, himself, acknowledged that the gain of the SVK-M was limited to about 30,000 because of this microseismic noise (outside of the frequency band of the signals we sought).

A Comparison of Seismograph Types

Growing exasperated with this argument for rejecting the Hardtack II data, I telephoned my associates in the United States and asked that comparison tests of the two instruments be conducted on an urgent basis. As a result, a Benioff seismograph and a simulated "Experts'" seismograph with peak response at 1 cps were operated side-by-side on the same pier in a vault at Ft. Sill, Oklahoma. Both response curves were measured with a sine-wave calibrator, which verified that the response of the simulated instrument was close to published response curves for the SVK-M, but with a reduced response to the six-second noise. The latter feature made it possible to operate the instrument at higher gain than the SVK-M in the short period range where the signals of interest were.

The two instruments went into operation on 9 December 1959. The gains were adjusted so that both seismographs had the same magnification at a 6 second period, the approximate peak in the microseismic spectrum. Nature cooperated by providing earthquakes, and several excellent teleseismic recordings of P-waves were made on 11 December. Copies were made and hand-carried to Geneva. On 15 December I presented copies of the seismograms and our conclusions to the British and Soviet delegations: the signals were far more clearly recorded by the Benioff than by the 1 cps instrument. Of most importance, the critical direction of first-motion of the P-waves was an order of magnitude clearer on the Benioff (see Figure 9.4).

Mikhail Sadovsky responded with congratulatory words on the "new instrument which I think would be superior to the present Benioff one. The first experiment has not been very good, but I do not think that this would necessarily determine lack of success of future experiments and recordings."

Fisk asked if the comment was intended as humor, to which Sadovsky responded, "I was serious." He went on to dismiss "an arbitrary interpretation of Dr. Romney's that the instrument you have put together does meet the requirements of the Experts." But as Dr. Panofsky pointed out, on the previous day Fedorov had identified a response curve essentially identical to the 1 cps instrument as conforming to the Experts' instrument, even though we had shown it to be inferior to those we had used for Hardtack II.

The matter was almost dropped at that point, but in the final report the Russians repeated their charge that "not one of the seismographs used conformed in parameters to the recommendations in the Experts' report." (They neglected to add that the performance of the seismographs we had used was much superior to the instrument that they claimed did conform.)

Magnitudes

Very early on in the meetings the Soviet scientists questioned the validity of the magnitude measurements at the Woody, Barrett, and Riverside stations -- the smallest measured magnitudes on *Blanca*. I explained that the observed scatter in the magnitudes was not unusual, and that the high and low values were simply manifestations of heterogeneities in the earth's crust. Sadovsky opined that the three stations were on unsatisfactory geological foundations. Frank Press, the Director of Caltech's Seismological Laboratory that operated the three stations, responded that two of the stations were on granite and the third on competent rock, and that "Woody is the best station we have." Sadovsky responded that he had thought that I had questioned the validity of the data (possibly as a result of a translation error). But the questioning continued and it became clear that the Soviet objective was to increase the average magnitude of *Rainier, Logan,* and *Blanca* by finding arguments to reject the lower individual magnitude measurements.

Dr. Yuri V. Riznichenko emerged as the primary voice of the Soviet side on magnitudes. In a lengthy talk during the third meeting, he asserted that our revision of *Rainier*'s magnitude was the cause of revised seismicity estimates. Our subsequent response that *Rainier*'s magnitude hardly mattered—the real cause was the first power slope of the magnitude vs. yield curve—fell upon deaf ears. Riznichenko chose to ignore the implications of the magnitudes of *Tamalpais* and *Neptune*; and would continue to find arguments to disqualify these shots from consideration throughout the remainder of the conference.

Riznichenko went on with a statistical discussion of the *Logan* and *Blanca* magnitude measurements. In what he called "Zone 1," within 1100 km from the shots, he reported a magnitude of 4.88 ± 0.24 for *Blanca*, excluding data from Woody and Barrett. He initially justified this with the comment that the measurements were "obviously omitted" as "extremely small in values." He would later argue that they should be excluded as lying outside the statistical scatter of the eight remaining measurements.

His "Zone 2" extended from 1100 to 2500 km. He characterized this as the "well-known shadow zone," although the Soviets had initially argued against its existence at the Conference of Experts, and both sides had agreed that strong P-waves through the mantle arrived well before 1,500 km. "In view of this, it is more sensible to leave these [magnitude] points out..." With this, he dismissed all of the lower magnitude (m_b) measurements within the teleseismic range starting at about 1700 km.

Turning to his "Third Zone," he gave us magnitude estimates from three Soviet stations. The stations were Tiksi, in the Siberian arctic, and two "temporary" stations. Tiksi, 6890 km from the explosions, had recorded both *Logan* and *Blanca*. The temporary stations—neither location identified -- had recorded *Blanca* only. (One temporary station at a distance of 8,300 km may well have been the classified Russian nuclear surveillance station we learned about decades later near Kuldar, on the mainland west of Sakhalin Island.) All three stations had recorded on SVK-M seismographs. So, of the 100 or more seismic stations in the USSR only one had recorded the 5 kt *Logan*, and only three had recorded the 19 kt *Blanca*! Rather a far cry from Pasechnik's assertion at the Experts' Conference that 5 kt would be readily detected worldwide. Furthermore, the largest of these four signals, *Blanca* recorded at Tiksi (Figure 9.5), appeared to begin with a rarefactional first motion! (And, it should be noted, on the instrument that Pasechnik claimed corresponded to the Experts' specifications.) Weak signals were also recorded at Mirny, a Soviet station in Antarctica, and Pruhonice, Czechoslovakia, but magnitudes were not determined from those stations. Apparently the Soviets had some calibration problems, too.

Combining the three Soviet measurements with those I had previously given, Riznichenko gave an average magnitude for *Blanca* of 5.117, contrasted with the magnitude 4.8 that I had reported based on all teleseismic data (except the new Soviet measurements which I had not seen, and that were of doubtful relevance because of instrument response differences which would tend to bias them higher than our measurements.

Riznichenko continued with an extensive statistical discussion, including the development of an elaborate scheme for weighting the various measurements. I might comment here that the talk was almost impossible to follow at the time. The translators were probably as baffled as we were, trying to cope with such intricate and highly technical subject matter. It was not until we had received a copy of the actual paper that we could understand what Riznichenko had done. We found that his weighting scheme put almost all the weight on the average of the few measurements beyond 2500 km. As a result he concluded that *Blanca's* magnitude was 5.1, *Logan's* was 4.8 and *Rainier* 4.6. These three estimates implied a magnitude vs. yield curve with a slope of 0.7, and far fewer numbers of equivalent earthquakes at low magnitudes than we had estimated.

Dr. John Tukey, a brilliant statistician from the Bell Telephone Laboratories and Princeton University challenged Riznichenko, pointing out large statistical errors and distortions in his probabilities.[28] His criticisms had no effect on Riznichenko. Professor Hans Bethe also joined the statistical argument, all to no avail.

Perhaps feeling somewhat challenged by the eminent Dr. Tukey's exposure of his statistical arguments, a few days later Riznichenko returned with a "physical" argument for rejecting data from his "Second Zone." He claimed to see a "trend" in Zone 2 measurements in which magnitude increased with distance, and he argued that all magnitude measurements in that zone should be rejected (see Figure 9.6). He attempted to support this with statistical arguments noting that the means and standard deviations were different than in other zones. And again, John Tukey and other members of the U.S. delegations argued against his interpretation, as well as his statistics, in vain.

Riznichenko's argument was that the magnitude of a seismic event is solely a characteristic of the source, and therefore does not depend on the distance. His "trend," in his opinion, indicated that the measurements did not reflect the true magnitude, but rather were indicative of some other physical phenomenon. He postulated that the waves were not actually P-waves, but were diffracted waves instead, and thus could not give an accurate indication of magnitude.

He chose to ignore implications of the undetected signals between 3,000 and 3,500 km, known to be smaller than signals at greater and lesser distances, and thus indicative of lower magnitudes. He also ignored the absence of signals at most of the numerous European and Soviet stations, and considered only measurements at the three Soviet stations previously mentioned — logically much larger than the average (undetectable) signals at the other stations on Soviet territory. Clearly, averaging such magnitudes would give a larger value than a more representative set of data.

In a series of papers published in 1956 Gutenberg and Richter had reviewed the relationships between the three magnitude scales, M_L, M_s and m_b, and their relationship to the energy of earthquakes. Their primary publication (Gutenberg and Richter, 1956a) pointed out:

> There are very few earthquakes for which one can determine M_s and m_b from amplitudes at distant stations, and, in addition, M_L, the "local magnitude," from torsion seismometers at short distances.[29] Most of our data for these are from the Kern County, California, earthquakes of 1952. These are used for Figure 10, which suggests that for near 6, approximately $m_b = M_L$.

The last sentence was based on only six earthquakes between magnitudes 5.6 and 6.2.

However, combining empirical relationships between m_b and M_s, with that between M_L and M_s, they inferred a relationship between m_b and M_L. (Gutenberg and Richter, 1956b).[30] They cautioned, however, that "the relationship of M_L to m_b is not yet on a definitive basis....."

Arguments about magnitude continued throughout the meetings. In the seventeenth meeting Riznichenko triumphantly announced his "discovery" of the "main and most serious error" in the U.S. working paper. He continued with a lengthy dissertation on the three magnitude scales, and then announced:

> A flagrant error in the working paper on new seismic data, an error continued in Dr. Romney's presentation and all other computations by the United States delegation regarding magnitudes lies in that, in the first zone... not the unified scale but the local M_L scale was used for the computation of magnitudes

He continued that "it was with very great difficulty that I was able to unearth this fact," (although it was clearly stated in the copy of my published paper presented to them at the second meeting, as well as in my presentation). He went on to outline the "unpardonable error" of not reducing all observation to a "single system of units."

He defended his own personal action in selecting only certain data, while rejecting other data, asserting "total averaging as a process which is the method of the blind and the lazy." "Any real specialist," he said, must "pick out only reliable data."

The referenced Gutenberg-Richter formula implies that m_b may be 0.4 to 0.7 magnitude units larger than M_L in the magnitude 4-5 range. We had not found such a difference to exist in the actual measurements of

Hardtack II. Our Hardtack II data had, for the first time, directly measured m_b and M_L for events smaller than magnitude five, and our measurements were numerically the same.

Although he had endorsed the relationship just mentioned in 1956, (as qualified in the previously referenced paper), in 1958 Charles Richter was apparently having second thoughts. In his paper in "Science" that year, Richter (1958) wrote, "However, neither [M_s or m_b] can now be related definitively to the values of magnitude as determined by the original method [M_L] ..." He continued with the caution that until such time as the local scale can be replaced by the teleseismic scales the use of the m_b scale [for local shocks] can only lead to error. I pointed this out to Riznichenko, who ignored it.

Undaunted by either our direct evidence, or Richter's caution, Riznichenko asserted that all of the M_L measurements should be "corrected" to m_b, using the Gutenberg-Richter [tentative] relationship. "Correcting," of course, meant adding 0.4 to 0.7 to all M_L measurements which would inflate the average magnitude. Not satisfied with this, he conjured up an elaborate set of statistical arguments for rejecting the two lowest values (at the Woody and stations). The essence of his argument, however, is that eliminating these two values reduces the dispersion (scatter) of the data. Seem obvious? Of course it is! Statistics applied to small sets of numbers can lead to strange results in my experience. Remembering Perry Byerly's "proof" that it is always dark in Kansas, we might ask if there is a physical reason to disallow the data. Quite the contrary—we have Frank Press' statement that the two stations are among the best of the Caltech network. We can also note that the range of the observations—3.9 to 5.4 for *Blanca* and 3.7 to 5.0 for *Logan*—is not unusual. Indeed, if the lower values are not included the range is unusually small.

In the end, Riznichenko included the contested data from Woody and Barrett (although given such low weights that they hardly mattered), but he continued to "correct" those M_L measurements to m_b, in spite of the cautionary words in the Gutenberg-Richter papers and Richter's later conclusion that neither M_s nor m_b can be related to M_L. The combined U.S. and Soviet measurements, he asserted, indicated that the magnitude for *Blanca* was 5.2 ± 0.1, and for *Logan* 4.7 ± 0.1, where the precision index is the standard deviation of a single observation. Any seismologist experienced in measuring magnitudes will recognize immediately that a standard deviation of a single observation as small as 0.1 magnitude units is most likely an indication of incomplete data, rather than an indication of high accuracy. Although unrecognized by its modern name at the time, we now know this was another example of magnitude bias, in this case caused in part by deliberately discarding lower magnitude measurements, and heavily weighting only those (larger than average) signals seen above the noise at great distances. And also using 20/20 hindsight, we now know that properly corrected, converting M_L values to m_b would have reduced the magnitude of the three shots rather than increased them (see the comment on M_L and m_b near the end of this chapter).

Revised Amplitude vs. Yield Scaling Law

On the second of December Professor Hans Bethe presented a theory relating amplitude of seismic waves to yields of underground explosions. Like the theory of Latter et. al. (1959), the theory predicted that amplitudes would be directly proportional to yield. On the seventh of December, however, he amended the theory. For explosions of *Rainier*'s yield or smaller, the theory should be correct, he concluded. For large explosions, however, his theory indicated that amplitude should increase as the 2/3 power of the yield. The yield at which the transition would take place could not be precisely calculated, but he estimated that explosions of 30 kt would be beyond the transition.

A curve having a first power slope at low yield and a 2/3 power slope above about 10 kt could be within the probable error limit of our observational data. It had little effect on the numbers of small earthquakes that the Control System would have to deal with. Changing the relationship between yield and magnitude also changed the relationship between yield and estimated number of shallow earthquakes. Dr. Panofsky presented the revised numbers to the conference, as shown in Table 9.1.

Decoupling

After an initial challenge to Fisk's proposal that the group hear a presentation on evasive methods for testing, Fedorov reluctantly agreed that it was within the scope of the working group's Terms of Reference. Hans Bethe had been asked to introduce the concept. Highly respected by the Soviet scientists, Bethe carefully laid out the basics of decoupling.

He was followed by the discoverer of the concept, Dr. Albert Latter. Latter explained the details of the theory. He presented an estimate that a large cavity in a *Rainier*-type medium (tuff) would reduce the signal by a factor of about 40, but that if the cavity were in a hard rock, like salt, the factor was estimated to be about 300 relative to a shot in tuff. Latter went on to describe the possibility of even further reductions in signal strength if the cavity were to contain an energy-absorbing material like carbon dust. Finally, he mentioned engineering opinions that it was feasible to construct cavities at least large enough to decouple 50-100 kt.

Soviet scientists expressed skepticism, but promised to study the ideas presented. It was only after several additional meetings, and frequent, pointed references to decoupling by Fisk, that Fedorov responded by announcing confidently that the Soviet side would show how and why they disagreed with Bethe and Latter.

Fedorov then called successively upon Drs. Gubkin, Sadovsky, and Pasechnik to speak. Gubkin presented complex theoretical arguments against the decoupling concept. It was difficult to follow, especially through an uncertain translation process. He did write a number of equations on the blackboard, however, which we could use later to try to reconstruct his arguments. His conclusion was clear enough: in his view there would be no difference between signals from well-tamped shots and shots in a large cavity.

Gubkin was followed by Sadovsky and Pasechnik, each of whom described experiments in which signals from well-tamped shots had been compared with signals from shots in cavities. Both reported *larger* signals from the cavity shots. Pasechnik speculated that the cavities caused the seismic spectrum to shift toward lower frequencies, to explain these measurements.

Upon analysis, Gubkin's equations proved to be equivalent to Latter's for seismic waves in a cavity in which the walls moved elastically. Thus, rather than refuting Latter's theory, they supported the theory, as Latter demonstrated at the next meeting using Gubkin's own equations. Furthermore, the experimental data of Sadovsky and Pasechnik were from explosions fired in chambers too small to meet Latter's pressure criteria, so were not relevant. Unfortunately, the U.S. did not have experimental data to validate the decoupling concept. Work was underway in the U.S. under the "Cowboy Project" to obtain such data, but it would not be available for a few months.[31]

However, the U.K. did have data demonstrating partial decoupling. It was obtained under their "Orpheus" project. Scientists of the U.K.'s Atomic Weapons Research Establishment had found an abandoned tunnel connecting two limestone quarries in Staffordshire, England. There, they had fired a chemical explosion of about 30 kg in a chamber a few meters in each dimension, created by blocking off the tunnel on either side. They found that the signals were somewhat smaller than those from a tamped explosion of about 3 kg. The data were presented by Dr. Ieuan Maddock, who reported that, using "only a crude chamber and using conditions which are a long way removed from those postulated by Professor Bethe and Dr. Latter, it is possible to achieve a decoupling factor of about one order of magnitude..." Maddock's experiment was not the definitive test that we would have liked, but the U.K. experiment clearly demonstrated that there are at least some conditions under which seismic waves from explosions can be reduced significantly.[32]

Full understanding of this rather esoteric concept came slowly to some on the Russian side, but it did finally come. It was a dramatic moment when the Russian scientists finally convinced Fedorov that decoupling, in principle, was a problem that must be faced. The report of the Conference of Experts was clearly inadequate as the source of all technical wisdom for the negotiations in at least this one important matter. My own impression at the time was that Fedorov literally turned gray in the face, and appeared to be taken suddenly ill. Nevertheless, the formal Soviet position, as summed up by Sadovsky, was "we really do not harbor any objections to the theoretical construction put forward..." but "... at the present time the decoupling theory is speculative and not confirmed by facts."

Improvements

Although the Russians continued to support Ambassador Tsarapkin's stance that no changes to the Conference of Experts' system were needed, they willingly agreed to support "possible improvements of the techniques and instrumentation." Most of these improvements originated from the American side, drawing heavily on recommendations of the Berkner Panel and work I had presented, demonstrating noise reduction at about 2 Hz proportional to the square root of the number of detectors in 3 km arrays of 4, 10 and 20 detectors. Concrete recommendations included installing 100 element arrays at all control posts (in lieu of the Experts' 10 element arrays), long-period seismographs at *all* control posts (in lieu of the Experts' "At certain posts"), and "The choice of the optimum short period seismometer as the component of the arrays recommended by the Conference of Experts for purposes of the registration of first motion in the presence of noise." The later, of course, reflected our demonstration of the superiority of the Benioff seismograph's response to that defined in the Experts' report (Russian version). There were also a number of other agreed recommendations, but these were described as preliminary or requiring further study. At the end of the meetings, these improvements became the only substantive matters to be included in the agreed portion of the final report.

Objective Instrument Readings

The attempt to define criteria for the initiation of on-site inspection brought fundamental differences into sharp focus. In the Soviet view, inspections should be rare events, occasioned only when compelling evidence of a nuclear explosion was present. In the Western view, such evidence was unlikely to be present in seismic signals. Accordingly, any seismic event not clearly shown to be an earthquake was potentially suspicious, and therefore should be *eligible* for on-site inspection, even though resources would limit those that could actually be inspected to relatively few.

The scientific spokesman for the Soviet side was Dr. A. I. Ustyumenko. He proposed to exclude suspicious events in densely populated areas and at depths "beyond possibilities open to mankind." Both, with more rigorous definitions, could be made to be acceptable to the western delegations, but he also proposed to exclude, as an earthquake, any event for which one or two rarefactions were recorded within 3,500 km. Furthermore, to be eligible for inspection he would require that the Rayleigh wave frequency be four times greater than those typical of earthquakes. Either of these latter conditions would have meant that *Blanca*, the largest underground explosion to date, would be excluded from inspection!

On the U.S. side, we struggled to formulate more precise rules that would not misidentify explosions as earthquakes. We excluded events deeper than 60 km as well as those that were located under deep, open oceans when not accompanied by hydroacoustic signals. The critical question, however, was how to specify conditions that allowed first motion to be used reliably. We knew that the apparent rarefactional first motions observed from explosions were just that—*apparent* first motions. In such cases the true compressional first motion of P was obscured by noise, leaving the much larger second half-cycle of motion to appear to be the beginning of the signal. But how could one be sure?

After measuring the first, second and third half-cycles on all Hardtack II signals that had unambiguous beginnings, we concluded that we could be certain if the first half cycle was twice the noise, and the second and third half-cycles were 20-40 times larger than the noise.

Although the Soviets presented no credible criteria for identifying an explosion, nor even causing an event to be deemed highly suspicious aside from a location in an aseismic area, Fedorov continued to insist that such criteria must be found. He castigated our attempts to eliminate earthquakes from suspicion, claiming:

...you want to provide the widest possible pretexts for inspection, but you are not worried about the nature of the pretexts. The essence of your scientific position—to call it that—is to give the greatest number of possibilities for inspection... whether they may be in search of alleged explosions or for some other purpose seems immaterial.

He went on to say there seemed to be a desire on our part to avoid the task set before us by the political conference.

The U.S. delegation also defined a number of kinds of auxiliary information, suggestive of earthquakes, but requiring further study or development before they would be relied upon as positive indicators. We referred to those features as "diagnostic aids," suggestive, but without the positive status of "criteria." In the end however, our set of criteria were ineffective except for large earthquakes.

Final Reports

As the meetings continued into mid-December, it became abundantly clear that fundamental differences divided the Soviets from the Western delegations. Arguments over the interpretation of magnitude data degenerated into not-so-polite wrangling. Fedorov's comments grew increasingly repetitive, long-winded, and hostile. In one accusatory outburst Fedorov asserted: "Why did you give the new seismic data? Was it to help a solution? No... the new data was only prepared to make the situation worse." Attempts by me and others to analyze the data in new ways to explain some point of view were greeted with charges of "manipulation" or "introducing new data" late in the meetings. The Soviets insisted that the working group *must* specify criteria that would remove a large fraction of seismic events from eligibility for inspection. "After all," said Fedorov, "the purpose we have as scientists is to try to help our political officers to identify scientific events." To which Fisk responded: "Science is not the servant of political expediency and we must not attempt to make it so." Fedorov countered with extended comments on the duty of scientists to serve the needs of the state, which included a bitter personal attack on Al Latter. A break-up without an agreed report became inevitable.

On 19 December 1959 TWG-II reported to the Conference that it had agreed on possible improvements to instrumentation and techniques for the control system, but disagreed on other matters discussed during their 21 meetings. A short agreed report containing little more than the agenda and an "Annex I" that described improvements was presented to the Conference, along with separate annexes prepared by each of the three delegations.

The Soviet report was delivered to the Conference by Fedorov, along with a vicious personal attack against the entire American group, and especially our leader, Jim Fisk. He accused us of manipulating data, of unscientific attitudes and of subservience to politicians.

Fisk responded:

Since Dr. Fedorov has read his incorrect, distorted and misleading statement, I feel that the record would be lopsided if I did not make a few moderate comments on behalf of the United States delegation.

He continued with other "moderate comments" directed primarily to the technical aspects of the Soviet report.

The Soviet Annex

Annex II "Statement by the Soviet Experts" repeated the now all too familiar criticisms of the new seismic data, complained that they had insufficient time to analyze the large number of seismograms we had given them, and that the magnitudes contained in the U.S. working paper of 5 January 1959 were "meaningless." They claimed to have uncovered "many errors.... and even some misrepresentation "in U.S. documents and statements, concluding that we had made "tendentious use of one-sidedly developed material for the purpose of undermining confidence in the control system...."

In the end, the Russian delegation, using Riznichenko's arguments, reported that we had provided "meaningless" magnitudes as a results of "erroneous operations" and they concluded:

> The Soviet experts did a more careful analysis of the seismic data on the *Blanca* and *Logan* explosions as well as for *Rainier.* This analysis was based on the well-known work on magnitudes of earthquakes and explosions done by Gutenberg and Richter between 1956 and 1958. As primary data the Soviet experts used all the magnitudes in the first (M_L) and third (m) zones (i.e. exclusive of the shadow zone) which had been presented by the United States delegation at the meeting on 14 December. All of the magnitudes were reduced to the unified scale of m.
>
> As a result, magnitudes for the explosions were arrived at as shown in Table 9.2.

The table gave U.S. magnitude for *Rainier* and the Hardtack shots, and "corrected" magnitudes according to Riznichenko. The report went on to associate the inflated magnitudes with Gutenberg's earthquake recurrence numbers (see Table 9.3).

> Thus, on the basis of a more careful analysis of the new seismic data, the Soviet experts have come to the conclusion that the annual numbers of earthquakes throughout the world equivalent

to explosions of given yield are, if anything, smaller than the numbers estimated at Geneva in 1958 and not 1.5 or 2 times greater, as is asserted in the United States documents.

For computing the continental earthquakes only, the numbers in the above table must be reduced by a factor of 2.

Although the Russians used Riznichenko's magnitudes to claim reduced numbers of equivalent earthquakes, there is room to doubt that even they took the increased magnitudes seriously. Larger, more detectable, signals would have implications for the control system itself. For example, the number of control posts could be reduced, or the arrays could actually be decreased in size, rather than increased. They made no mention of such implications.

Decoupling was dismissed with:

During discussion of the amplitude of the seismic signal produced by an underground nuclear explosion, the United States delegation introduced the idea that the seismic signal could be disguised by carrying out the explosion in a sufficiently deep and large underground cavity. The United States experts' views on the possibility of considerably reducing the seismic signal under such conditions are, in the main, based on general considerations of theory. However, even from the theoretical point of view the earth's crust is a very complex medium. Therefore, a combination of formal mathematical solutions for the problems involved in the dissimilation of underground explosions does not as yet offer a sound basis for any findings relating to the possible amplitude of the seismic signal generated by an explosion in a deep underground cavity or to the technical feasibility of carrying out vast underground construction operations at a depth of the order of one kilometer.

Concerning "objective instruments readings," the Soviets gave no criteria of their own in their final report. Probably they recognized that the criteria proposed in the meetings by Ustyumenko were easily shown to be ridiculous. However, they commented on the United States criteria:

...the United States experts suggest a system of criteria which virtually rejects the very idea of selecting suspicious events from the events recorded by the control system

As a result of analysis, it became clear, for example, that if the United States criteria were used the determination of first motion could be considered basically reliable only in the case of explosions with yields of the order of <u>hundreds or thousands of kilotons</u>. Obviously, such a formulation of criteria is challenged even by the United States scientists themselves, their criteria would leave under suspicion the overwhelming majority of the earthquakes registered by the control system.

The Soviet experts submit that here their United States colleagues are on the brink of absurdity.

The British Annex

In Annex III, the U.K. delegation concluded that the Hardtack II data "have shown that the 1958 Experts were too optimistic" about the ability to identify earthquakes. They refuted the Soviet criticism that the instruments used to record Hardtack II explosions did not conform to the specifications of the Conference of Experts, pointing out that it was not "inferior for the registration of first motion in the presence of noise." Regarding decoupling, the U.K. endorsed the theory and reported that an order of magnitude reduction in seismological strength had already been demonstrated; however, they expressed uncertainties about the practical aspects of decoupling nuclear explosions. On criteria for on-site inspection, their view was that the Soviet-proposed criteria would correctly classify explosions of several tens of kilotons, but would misidentify smaller explosions as earthquakes. They concluded that "the best that can be done in the present state of knowledge is to define criteria which will identify a modest proportion of earthquakes"; the remainder must be regarded as eligible for inspection, even though the number will be too large for all to be inspected.

The U.S. Annex

Annex IV reiterated our conclusion that the amplitude of the first motion was smaller than believed at the time of the Experts Conference, and that numerous apparent first motions corresponding to rarefactions had been seen on the Hardtack II recordings. We noted that the direction of first motion was "unreliable in the 'shadow zone' that was shown by Hardtack data to extend from about 1100 to 2500 kilometers, instead of 1000 to 2000 kilometers as was concluded at the Conference of Experts." Earthquakes equivalent to explosions of 1 kt or greater were estimated to number about 15,000 annually. Decoupling by a factor of 300 or more relative to *Rainier* was possible in our view, and we reported that engineering studies indicated the feasibility of constructing cavities large enough to decouple explosions "at least as large as 70 kilotons." We listed a number of techniques and equipment with potential to improve the seismic system. For identification of earthquakes we proposed finding a focal depth greater than 60 km; occurrence under the deep, open ocean without generating hydroacoustic waves; establishing that the event was a foreshock of a magnitude 6 or greater earthquake (identified as such by other criteria); and the use of first motion (under strict conditions to avoid misidentification).

Annex IV also had an Appendix that was essentially a commentary on the results of the meetings. A key point addressed the problem of identification of seismic events:

> ...such criteria must be formulated so that a large number of explosions would not be classified as natural earthquakes and that the criteria must be based on well established technical information. Unfortunately, the resulting criteria classify only a small fraction of seismic events as natural earthquakes leaving a large number eligible for inspection...

The Appendix continued to point out that the Soviet-proposed criteria would classify explosions up to 19 kilotons as natural earthquakes. It concluded:

> In the view of the U.S. Delegation, the problem of the formulation of criteria is a strictly technical problem. If technical knowledge permits one to identify a large fraction of seismic events as earthquakes then it is clearly a great advantage to the control system. If technical knowledge does not permit this, then seismic events must remain eligible for inspection.

It had been a frustrating experience for the U.S. group, and also a bit disillusioning to find that our Soviet colleagues would accept such a patently political role. We had expected them to seek the most favorable interpretation possible, as they had at the Experts' Conference, but we had also expected them to join with us in offering our governments our best scientific advice. Instead, the political conference was left deadlocked.

A Comment on First Motion

The relative weakness of the first compressional motion in the P-waves had been an unexpected feature of the Hardtack II signals. We had noted that the amplitude of the first motion decreased more rapidly with distance than did the main P-pulse. An additional, and disappointing characteristic of first motion from *Logan* and *Blanca* was that it was smaller than expected *relative* to the half-cycle that followed. Our expectations had been strongly colored by observations of first motion from near-surface air bursts. For example, the Hardtack II explosion *Socorro*, a 6 kt shot on a 60 m tower, produced typical first motions more than 1/2 the amplitude of the following, rarefactional, motion. For *Logan* and *Blanca*, on the other hand, the first motion was typically less than one fourth of the amplitude of the following motion. Since these observations were made only days apart, at the same stations, and on the same instruments, it was clear that this was a source effect. Werth et. al. (1962) later showed by modeling that the surface reflections from these relatively deeply buried events probably enhanced the following rarefactional motion. The implications, however, were clear whatever the cause: to be certain that we didn't misread the true direction of the first motion, the ratio of the peak signal to the noise would need to be substantially greater than we had earlier assumed. This, in turn, meant that our principal identification criterion was less effective than we had earlier estimated.

Another unexpected feature of the Hardtack II signals was the existence of precursory motion before the main P-wave pulse at distances beyond about 1,200 km. The travel times, and apparent velocities of the main pulse led us to believe they were normal P-waves through the earth's mantle. The precursory motion, then, might be diffracted energy or, in some cases, Pn waves arriving before P. Whatever the causes, they complicated the determination of the direction of first motion in what we called a "shadow zone" between about 1100 and 2500 km. (A shadow zone for first motion, perhaps, but surely not a shadow zone for the strong P-waves observed in this zone. In hindsight, it was an unfortunate choice of words.)

Two years later, Inge Lehmann (1962) conducted a thorough reanalysis of travel times from *Logan* and *Blanca*. To explain the observations, she developed a crustal and upper mantle model containing a low velocity layer at depths between 150 and 215 km. The agreement between her calculated travel time curves and the observed data provides strong confirmation that the large amplitude, late arrivals beyond about 1100 km are the teleseismic P-waves through the deep mantle of the earth. At distances less than about 1700 km, they may be preceded by weaker waves (that we now call Pn) traveling through the uppermost mantle above the low velocity layers and interfering with the determination of the direction of first motion. In her interpretation, the true Pn waves traveling at the base of the crust died out at distances near 600 km.

A Comment on Response Curves

In 1970 Pasechnik published what was to become the "bible" for nuclear test detection seismologists of the Soviet Union. The book, "Characteristics of

Seismic Waves in Nuclear Explosions and Earthquakes" (Pasechnik, 1970), is comprehensive in scope, beginning with a description of equipment used for recording explosions and continuing with descriptions of the waves recorded at various distances. When next we met (in 1971), Pasechnik presented me with a suitably inscribed and signed copy of the book.

The book contained numerous tables, copies of seismograms and figures. Glancing through, I could recognize names and dates of familiar nuclear explosions, measured amplitudes and periods, etc., although my knowledge of the Russian language is almost negligible. His Figure 2 caught my eye. It displayed easily recognizable response curves for four seismographs: the Benioff short-period and the Sprengnether long-period instruments made in the U.S. and the Soviet made SK and SK-M instruments. Translated, the caption reads "Amplification curves of seismic equipment with the parameters recommended by the Conference of Experts of 1958. Equipment installed in stations for detecting nuclear explosions of the United States and Other Countries." The Benioff response curve is identified as "array of 10 short-period seismographs." So, a dozen years after the acrimony of Technical Working Group II over this very point, Pasechnik acknowledged that the instrument we had used to record the Hardtack II explosion did, in fact, meet the specifications of the Conference of Experts!

A Comment on M_L and m_b

Measurements of seismic waves from *Logan* and *Blanca* had indicated that the "local" magnitude M_L was very nearly equal to the body wave magnitude, m_b. This was somewhat unexpected, since the equation of Gutenberg and Richter implied differences of the order of one-half magnitude unit. However, since we had measured both M_L and m_b on only two events, the data were hardly conclusive even though they strongly suggested that the Gutenberg and Richter formula did not apply in this magnitude range. As previously noted, Riznichenko had seized upon the tentative equation of Gutenberg and Richter and used it to inflate the magnitude estimates on *Rainier, Logan,* and *Blanca.*

It was not until 1972 that a better substantiated relationship was published. Few regions of the earth have good networks of Wood Anderson torsion seismographs (required for measuring M_L). Furthermore, good capabilities for obtaining m_b measurements on low magnitude events only developed in the mid-1960's, thanks to the World Wide Standard Seismic Network and their sensitive short-period Beniofff seismographs. In New Zealand, Gibowicz (1972) had good access to both types of measurements. He obtained good quality measurements of both M_L and m_b on 69 shallow earthquakes in New Zealand with magnitudes between 4.6 and 7.3.

Data on m_b were largely taken from USC&GS reports, and reports of the International Seismological Center. Data on M_L was obtained from his local network.

He found, by the method of least squares, a linear relationship between the two quantities. He pointed out that the slope of the line was very nearly 1 and poorly determined by the data, and that "it could equally-well be assumed" that:

$$m_b = M_L - 0.45 \pm 0.03$$

A similar relationship was later found by Chung and Bermreuter (1981) for earthquakes in the Western United States. They concluded:

$$m_b = 0.99\ M_L - 0.39$$

Thus, instead of *increasing* M_L values to infer m_b, it appears Riznichenko should have *decreased* M_L values. As previously pointed out, there is no good theoretical basis for assuming that the relationship found from earthquakes necessarily applies to explosions. However, clearly Riznichenko's "corrections" were far too large, and probably even of the wrong sign.

A Comment on "Second Zone" and "Third Zone" Magnitudes

I have commented previously about Riznichenko's failure to consider the low amplitudes at stations between 3,000-3,500 kilometers where signals were not detected. While magnitudes cannot be measured in the absence of signals, it is possible to place an upper bound on what the magnitude could be based on the noise levels.

At the station at 3017 km, *Blanca* was recorded with a signal of 32 millimicrons, but *Logan* was not. The measured noise, 5-10 millimicrons, implies a *Logan* signal 1/3 or less than the *Blanca* signal, or at least 0.5 magnitude units smaller. This implies a magnitude of about 4.5 or less. The stations with similar noise and undetectable signals at 3300 and 3502 km may have had signals near the same magnitude, especially if the noise was nearer to 5 millimicrons. If these data "points," or rather upper limits, were also plotted on Figure 9.6, the seeming fit to Riznichenko's arbitrary and physically unsupportable curves would be substantially less suggestive.

But more fundamentally, dismissing the strong P-waves found in the second zone as diffracted waves is belied by the amplitudes themselves. True enough, on the U.S. side, we were concerned that precursory waves, possibly diffracted, interfered with detecting the true first motion, and we argued that first motion was unreliable in this zone. But far from being weak, as diffracted waves would surely be, the main pulse was clearly a strong, normal P-wave through the earth's mantle. As Figure 7.2 shows the true "shadow zone" is near 1,000 km, and much stronger normal signals arrived a few hundred km beyond. Furthermore, careful scrutiny of

Riznichenko's "trend line" will reveal that the data points do not actually fit the curve very well—a number lying a half-magnitude or more above or below the lines.

Clear exposure of the flimsiness of Riznichenko's "analysis" came several years later when larger underground explosions took place, and more stations observed them. The *Fore* explosion in January of 1964 was well recorded in the "second zone" in which Riznichenko attempted to dismiss *the low magnitude* readings. As seen in Figure 9.7, there are both low and high magnitudes in this zone, 1700-2500 km from the source. Thus the "trend" that was claimed to be the basis for rejecting measurements in this zone does not exist as a physically based phenomenon—rather, it is a statistical anomaly. Furthermore at greater distances the scatter about the mean magnitude is almost a full magnitude unit (as is normal for such events), and not the ± 0.1 magnitude unit that Riznichenko found from his biased data set.

I draw a lesson from this. It may be tempting to draw lines through or connecting points in sparse and scattered data sets. However, unless there is a sound physical basis for doing so, the lines may be misleading. It may be helpful in remembering this to think of famed astronomer Giovanni Schiaparelli, and his discovery in 1877 of "canals" on Mars. These found a place in astronomy textbooks for nearly a century afterwards, until exposed as figments of his imagination by photographs from modern Martian probes.

TEN

The Seismic Research
Program Advisory Group

TWG-II had ended in acrimony and frustration for the U.S scientists. Nothing had been accomplished. Neither the Soviet objective of codifying the seismic evidence calling for onsite inspection, nor the U.S. hope that presenting a scientific problem would lead to an objective search for a scientific solution. After a full year of negotiating, the two sides seemed far from a treaty.

Eisenhower's Reaction to TWG-II

The Conference on the Discontinuance of Nuclear Weapons Tests recessed on 19 December 1959 after receiving the separate reports of Technical Working Group II, and the scientists returned to their respective countries to brief their political leaders. President Eisenhower was reported to be angry at the outcome, and he castigated the behavior of the Soviets in a White House press release on 29 December:

> The prospects for such an agreement on a nuclear test ban have been injured by the recent unwillingness on the part of the politically guided Soviet experts to give serious scientific consideration to the effectiveness of seismic techniques for the detection of underground nuclear explosions. Indeed the atmosphere of the talks has been clouded by the intemperate and technically unsupportable Soviet annex to the report of the technical experts.

Nevertheless, he announced that the U.S. was willing to continue the political conference, even though no agreement was in sight. He noted that the U.S. voluntary moratorium on nuclear testing would expire on 31 December, and that he therefore considered the U.S. free to resume nuclear weapons testing. However, the President stated that the U.S. would not do so without announcing its intention in advance.

Funding for Research

Seven million dollars were provided to AFTAC on 7 January 1960 for seismic research and development, greatly increasing the $460,000 provided to us in the preceding October. The timing, given that we had first proposed our program during the preceding July, suggests that action had finally been stimulated by the outcome of TWG-II. The additional money came, as anyone familiar with bureaucracy might anticipate, with a requirement for a new program plan, which we prepared and submitted on 21 January to the Advanced Research Projects Agency (ARPA), the source of the funds. The program was structured under six major tasks:
1) Worldwide Standard Seismological Network (a project to be managed by the U.S. Coast and Geodetic Survey to re-equip U.S. and foreign stations with modern seismographs)
2) Generation and Propagation of Seismic Waves (largely basic university research)
3) Detection Methods (array design, ocean-bottom detectors, deep-well detectors, etc.)
4) Systems Development (design and test of "control posts," etc.)
5) Large explosions (for research on seismic effects)
6) Special Studies (Network capabilities estimates, etc. to support negotiations)

The program was soon approved, and contracts to begin work were written on an expedited basis.

Inspection Quotas and a Magnitude Threshold

Treaty negotiations resumed in mid-January 1960, and Soviet Ambassador Tsarapkin promptly called for agreement on the earlier Soviet proposal for an annual quota of on-site inspections. U.S. Ambassador Wadsworth did not respond directly, but instead pointed out the seriousness of the technical impasse stemming from disagreements in Technical Working Group II. He reported that the U.S. could no longer accept the report of the Conference of Experts as an adequate basis for estimating the number of seismic events that might warrant on-site inspection. He criticized the Soviet approach which "presupposes that science will always be the servant of politics," and stressed the need to establish an agreed scientific basis for initiating inspections. He urged the Soviets to give careful study to the criteria proposed by the U.S. delegation to TWG-II. Amid charges and counter-charges about national motives, as well as the motives of the scientists during TWG-II, the U.S. continued to press for a technical method of selecting events for inspection. The number of inspections must be linked to the number of unidentified seismic events, in the U.S. view. The Soviet Union, on the other hand, continued to press for a political solution — a negotiated quota.

In Washington, among members of the Committee of Principals, there was doubt that a comprehensive test ban treaty could ever be negotiated with the USSR. As a result, earlier suggestions that the United States attempt to negotiate a ban on underground tests larger than some threshold were revived for reconsideration (Appleby, 1957). George Kistiakowsky, then the President's science advisor, and thus one of the Principals, proposed a threshold defined in terms of seismic magnitude to avoid the arguments of TWG-II on the relationship between magnitude and yield (Kistiakowsky, 1976). The matter was reviewed by a panel of seismologists and other experts, and we concluded that the concept was feasible and that magnitude 4.75 was a suitable level (about 20 kt under *Rainier* coupling conditions).

Consequently, on 11 February, 1960, the U.S. made a new proposal for a phased treaty. In his introduction of the concept to the Conference, Ambassador Wadsworth made it clear that the U.S. would much prefer a cessation of nuclear tests in all environments. However, the failure of TWG-II left the Conference at an impasse on control of underground explosions, and, the U.S. could not agree to prohibit testing under conditions where controls could not be effectively maintained. The main elements of the proposal were to ban tests above ground to the maximum altitude of effective controls; to ban all underwater tests; and to ban tests underground of magnitude 4.75 or greater (about 15-20 kt under *Rainier*/Hardtack II coupling conditions). Proposing a threshold in terms of magnitude, it was explained, should circumvent technical disagreements between Western and Soviet scientist about criteria for identifying earthquakes, as well as about equivalent kiloton yields of events.

Ambassador Wadsworth also announced that the U.S. had already embarked on a major program of research and experimentation to improve seismic detection (the Vela program) and he invited the U.K. and USSR to join with the U.S. in this program. As the research produced more effective monitoring means, Wadsworth offered, the magnitude 4.75 threshold could be lowered. He proposed that the U.S. criteria of TWG-II be applied to events above magnitude 4.75, and that 30 percent of those unidentified would be selected for inspection.

After severe questioning, Ambassador Tsarapkin initially rejected the proposal, accusing the U.S. of merely seeking a way to resume testing. He was critical of expressing the threshold in terms of magnitude, arguing that TWG-II had demonstrated that the determination of magnitude was a controversial matter. He also charged that a control system was not needed for seismic events above the threshold since existing national seismic systems were adequate. However, several days later he began presenting counterproposals to the Conference modifying various aspects of the U.S. proposal rather than rejecting it outright.

Joint Seismic Research Proposed in Negotiations

Both sides compromised, and by the end of March 1960 there was agreement in principle on a phased treaty. It would ban underwater and atmospheric tests; it established a magnitude 4.75 threshold for underground tests; there would be an annual quota of on-site inspections; and there would be a moratorium on tests below magnitude 4.75 during a period of joint research and development to improve seismic capabilities. However, the Soviet agreement had several conditions attached to it: the annual quota applied to tests below the threshold as well as above; and the quota must be arrived at by political means, that is by negotiation, and not as a percentage of unidentified seismic events. In the Soviet view the research, and thus the moratorium, should extend over a period of four or five years. The joint research would include a "strictly limited number of [nuclear] tests," and it was tentatively agreed that a meeting among scientists to plan the program would begin in mid-May, 1960.

Although the emerging agreement fell short of the 11 February proposal by the U.S., and the Soviets had attached conditions that would require difficult negotiation, the Soviet position was generally regarded within the U.S. Committee of Principals as a great advancement over their earlier positions. The pessimism of the winter was replaced with optimism within the Committee that a treaty would be negotiated after all, with the further benefit that the ensuing activities of international inspectors would gradually "open up" the Soviet Union and help break down their traditional secrecy.

U.S Preparations for Technical Meetings

As the treaty negotiations advanced, and it became evident that a coordinated or joint research program might become a reality, our AFTAC/ARPA program became of increasing interest to U.S. policy makers. It was, after all, the major element of the research program that the West had offered to conduct in cooperation with the USSR and U.K. As a consequence, a "Panel on Seismic Research and Development" was convened by Dr. Kistiakowsky. The panel was chaired by Dr. Wolfgang Panofsky of Stanford University, who had chaired TWG-I and been Deputy Chairman of TWG-II.[33] The Panel met in the old Executive Office Building on 7 April 1960, to "review the proposed seismic R & D program to be conducted in the United States in coordination with the U.K. and USSR." I recall quick agreement with most of our proposed research plan, the focus remaining primarily on the more controversial experiments involving large chemical and nuclear explosions. Kistiakowsky (1976) wrote of it: "Meeting didn't go well, soon developing into an interminable debate between people (mainly Dick Latter and General Starbird) who wanted to push nuclear tests and those who questioned their usefulness."

Certainly we all knew that the proposed nuclear explosions for seismic calibration, first recommended by the Berkner Panel, were the politically sensitive parts of the program. For this reason each shot needed to be challenged, and well justified. And although there might be differences of opinion on the number and objectives of the tests that were required, the scientists on the panel were unanimous on the need for at least some nuclear tests to achieve our research goals.

Several Panel members worked late into the evening and Dr. Kistiakowsky joined us for dinner. I remember the dinner in a nearby restaurant with the President's science advisor quite well, but recall few specifics of the conversation. I do recall Kistiakowsky's mention of a report under review by another panel that had studied the relative concentration of carcinogens in distilled spirits. The results indicated a direct relationship between the depth of color of the liquor and the presence of cancer-causing agents, a multiple filtered vodka being the safest. This was considered by the White House reviewers to be an extremely sensitive result that, when and if released, would cause strong reactions within the U.S. bourbon industry, and throughout the U.S. corn belt. I gained a strong impression that politics could, at times, be at least as important as science to White House staff.

Our panel met for a second time on 13 April at Stanford University. The final report, concurred in by all members of the Panel, was completed in this meeting. In addition to the technical aspects of the proposed U.S. program, the Panel also addressed "the extent of the coordination to be undertaken by the U.S., U.K., and USSR" and attempted to evaluate "the general results to be expected from the program during the next one or two years."

On the question of whether the program would be a "joint" program or a "coordinated" program, the Panel noted that the national program must be independent if results were to be produced during the next one or two years. An unreported, but obvious, reservation was that we didn't want to work under circumstances where the Soviets could veto elements that we thought were vital to our program, especially nuclear tests for research. We knew from Tsarapkin's words in the Conference that such tests had been accepted, only reluctantly, by the USSR. The Panel did point out that there were possible joint projects, and many areas where close cooperation would be beneficial.

On the technical aspects, the Panel described our plan as basically good, but felt that much of it would require more than two years. They also recommended that the program be broadened, and recommended several ways to do this, including adding a number of 100 ton chemical explosions, and getting laboratories of the petroleum exploration industry involved in the research. And they explicitly confirmed the need for nuclear explosions.

The predicted near-term results were caveated by pointing out that gains might be offset by more effective evasive methods, and that there is always a threshold below which any given system is ineffective. However, given sufficient priority and money (the Panel assumed $50 million per year — much more than I believed would be forthcoming), improvements of several kinds were possible. Looking back over my notes, the Panel certainly did not try to oversell the program. Although a number of promising detection methods, analytical techniques, identification methods, etc., were pointed to, the essence of the message to the President's science advisor was "we can't promise a breakthrough, but we can promise much greater knowledge about what seismology can and can't do for nuclear test detection." It was clear to all that our research would have to address some very intractable problems.

The U-2 Incident

On 1 May, as preparations continued for the research program meeting with the USSR and UK, Francis Gary Powers was shot down while flying a reconnaissance mission over the Soviet Union in a U-2 aircraft. Such missions had been going on since 1956, and had provided invaluable photographic information about the Soviet missile capabilities, and other military information unobtainable by other means from the secretive Soviet Union. Flying at altitudes above 70,000 feet, the U-2 seemed invulnerable until that time, in spite of repeated Soviet attempts to attack various flights over their territory.

When the airplane was first known to be missing, a "cover story" was released to the press in Washington, reporting that a U.S. weather research aircraft appeared to have wandered into Soviet airspace and was lost. Later, as more details were gradually released by the USSR, including the embarrassing fact that Gary Powers had been captured, the stories changed. Finally, on 7 May Premier Khrushchev announced that the Soviet Union had shot down a U.S. spy plane. We wondered if the incident would poison the atmosphere for the joint seismic research meetings but that appeared not to be the case.

Joint East-West Technical Meetings

At meetings of the Conference in Geneva in early May, Ambassador Tsarapkin announced the U.S.S.R.'s acceptance of an offer by President Eisenhower and Prime Minister MacMillan to declare a moratorium on tests below magnitude 4.75, and that a joint research program should be carried out. He conditioned this, however, with insistence that there be a "single joint program" rather than separate coordinated national programs as the U.S. and U.K. delegations had proposed. He also insisted that "the duration of the moratorium should not be less than the period during which the joint research program will be carried out."

The "Seismic Research Program Advisory Group" (SRPAG or "group" hereafter) convened in Geneva on 11 May 1960 as the treaty negotiators had agreed. The U.S. delegation was chaired by Dr. Frank Press; I served as his deputy.[34] The Soviet delegation was chaired by Dr. Mikhail Sadovsky, supported by the all too familiar Drs. Pasechnik, Riznichenko, Keilis-Borok, Ustyumenko, and others. The British delegation was chaired by Dr. Henry Hulme, a physicist from the U.K. Atomic Weapons Research Establishment. Hulme was supported by geophysicist Dr. Maurice Hill of Cambridge University, J.K. Wright of AWRE (who had attended TWG-II), and others.

The meetings opened on the warmest of notes -- a far cry from the bitter antagonism of our last meeting in Geneva five months before. In his opening remarks, Frank Press, the chairman for the day, welcomed our British and Russian colleagues with warm and hopeful words:

> I take great personal pleasure in renewing acquaintanceship with my colleagues from the United Kingdom and from the USSR. Much has happened since we last met, but now again a new scientific task brings us to Geneva. The circumstances of this Conference signify to me that at last we are moving forward on the problem before us.

Responses by Mikhail Sadovsky and Henry Hulme on behalf of the Soviet and British delegations were also cordial and positive about the work of the group.

Sadovsky immediately picked up on a reference by Frank Press to a press release by the political Conference defining our objectives as to assist "in considering the nature of, and making arrangements for, a program, or programs, of seismic research and experiment by the three countries." In

Sadovsky's view "it is better and easier to have a program agreed upon in all its details. Let us have a single program which, later, might be distributed or divided among the countries…" He was voicing a difference between delegations in the Conference, previously noted, as to whether the research would be "joint" or "coordinated." We had some concern about a "joint" program, especially if by this term the Soviets meant it should be subject to the "unanimity" (veto) principle. But it soon seemed to be a minor matter as we discussed areas of research that each of the three countries proposed to work on, and other areas unique to each country. As a practical matter it appeared to the U.S. side that we would have a "coordinated" program, perhaps with areas of "joint" research.

The U.S. Research Program

An introduction to the ongoing U.S. seismic research program had been given in Press' opening remarks; the program itself was described to the group in two parts. The first, a summary of the VELA underground program, was presented by Carlton Beyer. Beyer, an employee of the Institute for Defense Analysis, had recently been assigned to ARPA to head the VELA program, which by this time also included research on atmospheric, as well as high altitude and space, explosions.[35] He was not an expert in any of the scientific techniques for nuclear test detection, but was experienced in technical project management. He gave a twelve page written summary of the program to each participant at the meeting, and addressed the details. His presentation was largely in program management terms: project title, organization and dollar costs by fiscal year for each contract currently funded, or planned for the next fiscal year. The program was funded for only two years, as the Berkner Panel had proposed.[36]

I followed with a description of the scientific framework of the program, containing the objectives of the major tasks and, in general terms, our approach to achieving those objectives.

The questions that followed were all from the Soviet side, and they were directed to the scientific details. As expected, many of the initial questions were directed toward details of our nuclear explosion program and our instruments and deployment plans for making the seismic measurements. Often the questions implicitly or explicitly challenged the need for some of the nuclear explosions, or were critical of our plans for developing temporary seismic stations to record data from explosions. But the questions were phrased as if we were conducting a normal scientific meeting—consistent with the cordial tone of the meeting.

Sadovsky mentioned that he and his colleagues had difficulty directing questions to Beyer, whom they had not fully understood because "the characteristics of the program…were given in dollars." He continued "you say that $30,000 is earmarked for the determination of criteria. Does this represent a high or a small figure? This is unclear to us." He subsequently expanded this to say they would prefer a more "concrete outline of the individual tests," and "features concerning the instrumentation," rather than dollar costs.

As the meetings proceeded, we did provide more detail, but also explained that some of the requested information would emerge from the research itself, and thus could not be defined in "concrete" terms at the outset of our program. A few days later, Frank Press remarked, "It is difficult for us to gauge the extent of the Soviet research program. When we listed the expense of our program in dollars it must have confused our colleagues, but to us it indicated just how great our effort would be. If we could have some concept, perhaps in terms of number of man-years of effort, of the Soviet program, it would help a great deal."

The preceding is mentioned here to give the "flavor" of the obstacles we faced initially—cultural, rather than political or even scientific. The meetings continued on amicable terms.

U.K. and U.S.S.R. Research Programs

A program of research by the U.K. was introduced at the second meeting by Henry Hulme, followed by brief talks by Maurice Hill and John Wright elucidating some technical aspects. The British program was similar to the program we had proposed, except that it would be smaller and would include no large explosions. Hulme's talk included a statement of willingness to "cooperate in any program of nuclear and chemical explosions."

Two days later (on Saturday) Yuri Riznichenko tabled an "Annex to the Draft Program," a somewhat formal document apparently intended as a model for a final report by the group, defining objectives for the three countries, the duration of the program, etc., as well as specific tasks. It included "Field experiments with large explosions" for the Soviets, but concerning the use of nuclear explosions Riznichenko only said "this question would be dealt with later. In any case, it seems obvious to us at the present time that a certain number of coordinated nuclear explosions of definite magnitude and energy will have to be carried out by us."

From our perspective, the most interesting part of the Soviet program was where and how they would carry out large explosions. Riznichenko's proposal was that explosions and measurements be conducted along a single profile extending in a generally southwest to northeast direction in the southern border region of the Soviet Union, essentially from Stalingrad to Lake Baikal. Under questioning, Sadovsky subsequently described several already-planned industrial explosions ranging in yield from 500 to 3500 tons that would also be exploited as seismic sources.

Under questioning about whether the Soviets would work on methods for detecting decoupled explosions, Sadovsky gave what he termed a political response: "It would be very difficult for us to explain to our people why, instead of working out methods for the detection and identification of explosions, we wished to spend funds in order to work out methods for the secret carrying out of nuclear explosions." Not quite an answer to our question, but clear enough about a more interesting question. He subsequently also clarified Riznichenko's prior statement, by adding that the USSR would conduct no nuclear explosions themselves for research, but they would carry out measurements of US research explosions.

On the following Monday morning, 16 May, the SRPAG met to hear responses to various questions deferred from the Saturday meeting. There were lengthy discussions of the Soviet large chemical explosions and on expected differences in seismic effects from chemical and nuclear explosions. I attempted to enlist Soviet cooperation in our plan to equip the world's seismic stations with modern seismic equipment (with no success). Frank Press, noting that there was much in common among our technical programs, attempted to get the group to begin to focus on how we would cooperate in the future. In short, it was a very normal, polite scientific meeting.

Collapse of the Summit Meeting in Paris

Things were different in Paris. A long-planned summit meeting, involving leaders of France, the Soviet Union, the United Kingdom and the United States convened on 16 May, only to hear an angry Premier Khrushchev demand that U-2 flights over the USSR be discontinued, and that those responsible be punished . President Eisenhower coolly responded that he had already ordered the flights to stop but he refused to go further. The Soviet Premier refused further participation, and the summit meeting collapsed. Both sides stated a few days later that this should have no effect on test ban treaty negotiations. We continued with our technical meetings.

The U-2 incident, having such major international impact, became a subject for discussion during casual moments before and after our meetings. The Soviet scientists were curious about the airplane and its capabilities but, of course, we could tell them nothing. They were also proud of their country's ability to (finally) shoot one down. In response to a somewhat boastful comment on this ability, Dick Latter responded, tongue in cheek, "Ah, but just wait until you face the U-3!"

Drafting a Coordinated Research Program

With three research programs on the table, it was time to put them together and to define methods for coordinating work among the three countries. The next day the UK delegation initiated this task, and tabled a "Summary of Planned Programs of USSR, USA, and UK and Possible Fields of Cooperation." The summary was divided into four parts and attempted to list, side by side, activities of each country and the form of cooperation that might take place. It highlighted the fact that there was much in common among the three programs.

The U.S. took the next step on 18 May. Drawing upon Riznichenko's more formally structured document as well as the UK draft, we included a general statement of objectives, reorganized the major tasks, and made an attempt to define the primary "fields of cooperation" that should be considered as well as to advance specific ways in which we might cooperate. We organized the work into five areas of research, and listed numerous specific tasks in each area. Our descriptions were designed to be general and were couched in language that we thought each country might accept. Decoupling, for example, was mentioned only as: "Model experiments and theoretical studies will be conducted to investigate the role of various properties of the medium, and of the design and dimensions of the shot chamber." This draft was accepted as the basis for a final report by the advisory group; further work on it was relegated to an informal drafting group, which met outside of the more formal meetings.

There was no lack of discussion in the formal meetings, however. Continuing a process of critiquing the separate programs, we were told that we planned too few temporary seismic stations for monitoring the explosions (we had planned to employ 20 stations), and that these stations should operate continuously between explosions to record earthquakes as well. Both criticisms were technically sound, but they would require large expenditures unavailable within our existing funds. Carlton Beyer referred both matters back to the Director of ARPA who responded a few days later that he would provide the necessary funds. Sadovsky pressed for a program duration of four to five years, arguing that the work outlined could not be accomplished in two years. Faced with the reality that our program was only funded for two years, and we lacked authority to commit our government beyond that period, we could only respond that we hoped for significant results in two years, and believed that such results would justify a continuation of research.

The Soviets were also critical of our planned use of the Nevada test site for many of our explosions. They argued that the explosions should be in more seismically active regions, and that explosions should be fired at both ends of our seismic measurement profiles. We responded that NTS had a sufficient variety of rock types to explore coupling effects, that it was a good place to study depth effects and the relative effectiveness of nuclear and chemical explosions, and that safety considerations also favored NTS. (Besides, can anyone imagine an underground nuclear explosion at the end of a profile extending from NTS to Vermont or Maine?)

We, in turn, offered criticisms of the Soviet Program. We pointed out that their emphasis on chemical and industrial explosions could lead to ambiguous results. In Frank Press' words, "After all, we are not trying to detect a chemical explosion, we are trying to detect a buried nuclear explosion." We also questioned their plans to make measurements along a single profile, pointing out that this might bias their results since propagation in other directions might be quite different.

Contention over Explosions

Mikhael Sadovsky was highly critical of our explosion program, and especially the proposed low yield nuclear explosions. Concerning these he voiced strong suspicion that they were, in reality, directed to developing tactical atomic weapons. He clearly suggested that the USSR's policy that "a strictly limited number of joint underground nuclear explosions" meant 6 or 7, and not the 12 proposed by the US. We explained again that our program required shots in different media, at different depths, at different locations and of different yields, and these studies had led to our carefully considered estimate of 12 nuclear shots.

Richard Latter carried much of the weight of theoretical discussions on explosions for the U.S. delegation, and "Gerry" Johnson similarly carried much of the weight of discussion on the operational and practical aspect of testing. In response to a question from Sadovsky, by far the major speaker on explosions for the Soviets, Latter stated that "From the point of view of [generating large] seismic signals, the most favorable medium is one like water." (Certainly our experience up to that time was that underwater explosions produce the largest seismic signals.) He went on to say "I do not know what the extremes of media are. I believe we can say that soft media such as tuff, while probably not one extreme, are more favorable for producing seismic signals than hard media, such as salt." Sadovsky, on the other hand, reported that he and Pasechnik "have always felt that this is not so...and our recent experience confirms our point of view." He alluded to a recent chemical explosion of 700 tons carried out in marble, and reported that "The amplitude in our chemical shot was somewhat greater than the amplitude which the United States had during [the 5 kt] *Logan*." Signals more than seven times larger than we would have expected? If true, it was contrary to the expectation of U.S.

physicists. Either the Soviets exaggerated the signal size, or U.S. physicists (and seismologists) had much to learn.

A Framework for Joint Research Emerges

Methods for measuring magnitudes also became a significant topic for the SRPAG to discuss. The U.S. had tabled a working paper on the subject in the political conference on the same day our group had convened. The U.S. had proposed adoption of Gutenberg's m_b scale, using measurements from short-period instruments, as the standard for verifying the magnitude 4.75 threshold. Measurement would be restricted to distances between 16 and 90 degrees from the explosion point in our proposal, to avoid the contentious issues of TWG-II. Pasechnik brought the working paper to our attention with a question on how our program might determine methods for measuring magnitudes at distances smaller than 16 degrees. At another point Riznichenko proposed that we attempt to develop an "energy" scale to replace magnitudes entirely. While endorsing the idea in a general way, Frank Press, recalling the contention on magnitudes in TWG-II, proposed that these questions might be referred to a working group, which was agreed. The SPRAG did, however, have to sit through a repeat of Riznichenko's TWG-II conclusions on magnitudes, backed on statistical matters by Dr. V.F. Pisarenko. We were no more convinced of Riznichenko's arguments in May than we had been the previous December, but consistent with the ambience of our meetings, did not attempt to refute him directly.

At the tenth meeting on 23 May Sadovsky commented:

> We have the impression that in the course of the last two weeks we have done some good work, and we note with pleasure that we have agreed on a far larger number of points than those on which we have different views.

At the next meeting, on 24 May 1960, Dr. Ustyumenko, who had chaired the subcommittee meeting earlier that morning drafting a final report, reported to the formal meeting:

> The drafting sub-committee considered the amendments introduced by our delegation, and we would like to note with satisfaction that our United States and United Kingdom colleagues have basically taken into consideration the proposals we made, and have introduced their drafts. I think as far as these drafts are concerned we have reached a general preliminary understanding.

It seemed that we were very close to an agreed report, even though the Soviet Scientists might deem it necessary to insert footnotes condemning our decoupling study or questioning the need for some of our small nuclear shots.

Soviet Withdrawal and Collapse of the Meeting

Conditions were dramatically different when we met the next day. The meeting was first postponed at Sadovsky's request. Then he privately informed our US and UK delegation leaders, Frank Press and Henry Hulme, that the Soviet delegation had been ordered to withdraw from the meetings in the aftermath of the collapse of the Paris summit meeting on 16 May. Sadovsky also mentioned that he had contacted influential officials in Moscow to request their assistance in reversing this decision since we were making so much progress. It could take several days, he said, to know the results of this attempted intercession. We met with foreboding on that Wednesday afternoon, 25 May, for a brief meeting. It seemed clear that Sadovsky was constrained to say very little beyond agreeing to meet again on the following Monday.

The answer to Sadovsky's appeal to Moscow was revealed to us on Friday 27 May in the meeting of the Political Conference, to which members of our group had been invited. Ambassador Tsarapkin started with a demand that the Western nations respond to his earlier questions on the duration of the testing moratorium. He then delivered a lengthy and one-sided review of events leading up to the present state of the negotiations, which included the statement that "the Soviet Union always recognized and still recognizes" that the Conference of Experts' recommendations can ensure effective control of underground nuclear explosions. He continued with accusations that the United States carried out the Hardtack II underground explosions "in order to fabricate data...to cast doubt on the recommendations of the Experts." He continued his polemic, finally reaching the point we were listening for—and repudiating his own scientists in the process—"since the Soviet Union has no doubts regarding the validity of the reports of the Geneva Experts of 1958, it sees no need for undertaking any research or experiments on its own territory." Tsarapkin went on to acknowledge that the U.S. intended to continue its program, but demanded that "it should provide for mandatory participation of the Soviet scientists in all research, from the preparation of an appropriate program to the carrying out of research and on-site experiments and the processing of the results."

In the absence of Frank Press, who had returned to the United States the previous week, I chaired the SRPAG meeting on Monday. The agenda, as had been agreed, was that we would conclude our discussions on the means for coordinating our research, and then determine how to conclude our meetings. I called first on Sadovsky to give his views. At times he read carefully from a prepared text—a rather unusual occurrence in our highly informal meetings. Citing basic differences in viewpoint on decoupling and the required number of nuclear tests, and dissatisfaction with the duration of our program, he proposed that each delegation should give an oral report to its representative in the Conference, providing its views on the issues that divided us, and relegating the matter of nuclear tests to the political conference. In short, there would be no agreed report on research and no need to consider means for coordination.

At Henry Hulme's recommendation, each delegation agreed that we should also report the large areas of agreement among us. With closing words from Sadovsky, Hulme and myself reflecting on how much we had accomplished and how much we had enjoyed meeting in such an agreeable manner, I adjourned the meeting.

We parted with personal farewells and private expressions of regret. I think each of us at the meeting had been looking forward hopefully toward a truly international cooperative research program to solve important scientific problems that divided East and West at the political level. On a purely personal level, working together in the cooperative environment of the meetings had cemented bonds of friendship that had formed. Certainly, when Pasechnik and Pisarenko visited the United States in early 1974, I was happy to arrange for extensive technical briefings on our research program, and was delighted to have them as dinner guests in my home together with my wife Barbara and younger daughter, Kim. Following dinner Victor Pisarenko informed us that he enjoyed singing at such times, and the remainder of the evening turned into a songfest, (with Ivan Pasechnik as our audience). Fortunately, Pisarenko knew many American folk songs, and I can still remember him singing "On zee Banks of zee Ohio" with gusto (and strong Russian accent). An evening or two later, we enjoyed taking our new-found friends to dinner at the (to them) notorious Watergate Apartments and to a concert at the nearby Kennedy Center conducted by Aaron Copeland. Somewhat to my surprise, both were quite familiar with Copeland's work, and expressed great pleasure at actually seeing and hearing him conduct his own music.

I should also note in closing that the Soviet criticisms of our research program were, in the main, beneficial. For example, without these criticisms, I doubt that American seismologists alone could have succeeded in doubling the size of what became known as the Long Range Seismic Measurements Program (LRSM)—our program to measure seismic effects of future underground nuclear explosions.

ELEVEN

Research Begins to Pay Off

Initiation of Research

The U-2 incident and the collapse of the SRPAG meetings in 1960 ended all hope of obtaining a negotiated treaty during the Eisenhower Administration. However, the Vela Program began moving forward rapidly. On 15 June 1960 a meeting was held at AFTAC to review the status of the planned *Lollipop* explosion, a 5 kt nuclear explosion in granite at the Nevada Test Site (NTS). It was intended as the first nuclear explosion for seismic research. Its objective was to measure seismic coupling differences between granite and tuff by comparing its signals with those from the 5 kt *Logan* explosion. Preparations were also going on at NTS for large chemical explosions to measure the expected difference in coupling between nuclear and chemical explosions.

Lollipop was never fired as such. As preparations continued, the Department of Defense constructed a number of nearby underground structures to test their vulnerability to a nuclear attack. By including such obvious military experiments, the explosion could never pass as a pure seismic research explosion in the eyes of the USSR. Simultaneously, in Geneva, Ambassador Tsarapkin was rejecting "safeguards" that the U.S. proposed to reassure the Soviets that our research explosions could not be used for weapons development.

Research and development was proceeding on a number of fronts. Funds had been provided to the U.S. Coast and Geodetic Survey to initiate procurement of equipment for stations of the Worldwide Standard Seismological Network and the selection of stations to be upgraded was underway. Research programs had been funded at nine universities on subjects ranging from seismic wave generating mechanisms to measuring characteristics of noise on the ocean floor. Industrial firms were investigating detection methods based on placing seismometers in deep wells to get away from near-surface noise, and optimized array designs. Federally funded research organizations were studying unmanned seismic station design and the effectiveness of identification criteria. An additional task beyond those of AFTAC's original plan had been added in the spring of 1960, to initiate research on on-site inspection. In short, a broad-based and diverse research program was underway, funded at $8.8 million.

A few months later, in October 1960, the first "Geneva Station" began operating at Fort Sill, Oklahoma at the site of the former AFTAC training school. The station opened with much fanfare, attended by notables from Washington, D.C. including Chairman of the AEC, John McCone. McCone took a keen interest in a seismogram on display to help explain the station's workings to visitors. It was a recording of a large, recent earthquake in Kamchatka. Asked by McCone to explain how he could tell it was an earthquake rather than an explosion, one of the station operators apparently confessed that he couldn't be sure, based on the limited information available to him, but if it was an explosion, it must have been in the megaton range. McCone returned to Washington impressed by the difficulty of distinguishing between earthquakes and underground nuclear explosions, as we soon learned.

The station was equipped with a ten-element array using Benioff seismometers in shallow-buried water-tight steel tanks, long-period seismometers of both Spengnether and Press-Ewing design, broad-band and narrow-band seismometers (simulated Soviet SVK and SVK-M seismometers), automatic processing film recorders (Develecorders), magnetic tape recorders and other state-of-the-art equipment. A more complete description of the station is in an article by Gudzin and Hamilton (1961). The "Geneva" term referred to the Geneva Conference of Experts, and the station was designed to test one of several ideas on array design consistent with the Experts' report. After several months of operation it was reported that the station had a detection threshold of about magnitude four at teleseismic distances—an excellent result for that time. We named the station the "Wichita Mountains Seismological Observatory," which set the pattern for naming similar array stations planned or under development.

The AEDS Retrofit Program

Promising ideas were being adopted simultaneously by AFTAC, for use in a "retrofit program" for improving the AEDS. Research on improving the signal-to-noise ratio by placing seismometers in deep bore-holes had been underway by AFTAC prior to the beginning of the Vela program, and by late 1960 an experimental sensor was in operation at a depth of 100 feet at the classified station near Encampment, Wyoming. Surveys had been completed at several stations to increase the number of detectors in the seismometer arrays, and advanced recording equipment similar to that at the Wichita Mountains "Geneva Station" was planned for each station. Three component long-period Sprengnether seismometers had been installed at most AEDS stations; while not proven yet, the promise of such seismometers to aid in discrimination between earthquakes and explosions had been clearly pointed out by Jack Oliver and Frank Press in Geneva and at meetings of the Berkner Panel.

Early in 1961, a decision was made to replace the large and heavy (more than 100 kg) Benioff seismometers with the newly designed Johnson-Matheson (J-M) short-period seismometers. The J-M seismometer had been developed at the National Bureau of Standards under AFTAC sponsorship, independent of the Vela program, to improve performance beyond the Benioff seismometers. They were smaller, lighter, more stable, more linear, and their response curve was designed to be flat with respect to ground velocity in the P-wave frequency range. This latter feature meant that measurements of the quantity a/t, critical to good magnitude estimates, were not affected by difficulties in estimating the period (t). Later that year the Laramie Analysis Center was closed, all analytical functions having been transferred earlier to AFTAC, thanks to the availability of the Zipagram system. This retrofit program would not be completed at most AEDS seismic stations until 1962, limited by the availability of funds and the sheer magnitude of this modernization program.

A New U.S. Administration

U.S. national elections were held in November 1960 with John F. Kennedy claiming victory as President over Richard Nixon by a narrow margin. The nuclear test ban had not been a major issue in deciding the election since both major candidates intended to follow President Eisenhower's initiative in seeking a treaty. Not that the test ban was ignored during the campaign. Amid speculation from intelligence "experts" that the Russians had tested nuclear devices secretly in violation of the agreed test moratorium, Vice President Nixon commented on the seismic disturbance recorded at Ft. Sill shortly before John McCone's visit in October. Reportedly briefed by John McCone, the Vice President said: "It might have been an earthquake. It might have been a large underground nuclear explosion. We have no way of knowing." (as reported in the *Washington Post* on 28 October 1960). Assurance that the event was not nuclear followed promptly from the USC&GS, Department of Defense, and others.

Negotiations in Geneva recessed in early December, and at the recently elected President's request remained in recess until March to allow time to reformulate U.S. policy. In his first State of the Union message in early 1961, the President declared his intention to "exhaust all reasonable opportunities to conclude an effective international agreement banning all tests—with effective international inspection and controls." He also initiated a new study of the associated technical problems by a panel chaired by Dr. James B. Fisk, Chairman of the U.S delegation, and including a number of veterans of the Conference of Experts, *e.g.*, Dr. Hans Bethe, Dr. Harold Brown, Dr. Richard Latter, Doyle Northrup, and Dr. Frank Press. And he appointed a new science advisor, Dr. Jerome B. Wiesner, head of MIT's electrical engineering department. Wiesner had been active in disarmament matters for several years as a member of PSAC. He would play a very active role in test-ban policy during the next several years.

The Fisk Panel report was a comprehensive review of the capabilities and limitations of the monitoring system described by the Conference of Experts—the "Geneva System," as it was commonly called at the time. The report could do little more than confirm its adequacy for oceanic and atmospheric tests, and point out problems for monitoring space and underground tests; it was premature to have major advances from the research that had been initiated only in the previous year. However, the review must have provided a useful introduction into the technical aspects of treaty monitoring for President Kennedy's new team.

Following President Eisenhower's lead, President Kennedy placed responsibility for developing nuclear test ban policy on the Committee of Principals. The Principals in the new administration included the heads of the five agencies originally represented: Secretary of State (Dean Rusk); Secretary of Defense (Robert S. McNamara); Chairman of the Atomic Energy Commission (Glenn T. Seaborg); Director of Central Intelligence (Allan Dulles); and Science Advisor (Jerome B. Wiesner). But President Kennedy enlarged the Committee to also include the Chairman of the Joint Chiefs of Staff, (General Lyman B. Lemnitzer), the Director of the U.S. Information Agency, (Edward R. Murrow), and the Special Assistant for National Security Affairs, (McGeorge Bundy).

To further emphasize the importance of the test ban and other disarmament questions, the new administration also drafted legislation to establish an independent agency devoted solely to these matters. This legislation was approved by the U.S. Congress, and the Arms Control and Disarmament Agency came into being in September 1961. Under its first Director, William C. Foster, the agency began to play an active role in test ban negotiations, as well as in influencing our research program.

Geneva Negotiations Resume

When test ban negotiations resumed on 21 March 1961, the new U.S. ambassador, appointed by President Kennedy, Arthur Dean, submitted new proposals that the President and his Principals had developed, hoping they would break the deadlock in the Geneva negotiations. From our perspective, the most important new proposals were that the U.S. seismic research program (and the moratorium on tests smaller than the magnitude 4.75 threshold) would be extended to at least three years (good news!); the U.S. would agree to reduce the number of control posts in the USSR from 21 to 19; and the U.S. would accept the Soviet proposed "safeguards" to prevent research explosions from being used for weapons development. To accomplish the latter, the President would request the Congress to amend the Atomic Energy Act to permit Soviet scientists to be given blueprints and to inspect the interior of nuclear devices intended for such use. There were other concessions as well, and Ambassador Dean described the new U.S. position as meeting the Soviet position "more than half way." Anticipating that the U.S. proposals would evoke counter proposals from the USSR, I was asked to join the U.S. delegation, along with several other scientists, when the proposals were presented to the Soviets.

The Soviet Union made no attempt to reciprocate. To the contrary, Ambassador Tsarapkin had, in fact, opened the meeting in which our concessions were offered, by presenting a major new stumbling block for the negotiations—a demand for a tripartite management council for the Control Commission. Under this arrangement the council, which would manage all of the monitoring functions provided for under the test ban treaty, would consist of a representative of the USSR, a representative of the West, and a representative from a neutral country. All decisions would have to be unanimous—that is, subject to a veto. This was known as the "Troika" proposal, an allusion to the Russian name for a vehicle (most famously a sled) drawn by three horses abreast. The proposal was seen as a deliberate attempt by the USSR to impede the negotiations. They had earlier proposed Troika management for the United Nations, and they knew it was unacceptable to both the U.S. and the U.K. In the absence of any significant response by the USSR to our new proposals, there was little for me to do, and I was released from the negotiations in Geneva on 31 March and returned to Washington.

The impasse on the Troika and other issues—numbers of inspections (the U.S. and U.K., proposed 20 annually on the territory of each country, the USSR proposed 2 or 3), number of control posts in the Soviet Union, (the U.S. and U.K. had reduced their proposed number from 21 to 19 but the USSR would accept only 15), the duration of the moratorium and a host of organizational and management issues—continued throughout the following summer amid deteriorating relations between East and West on political issues outside of the Test Ban negotiations. As a notorious example, after refusal of the Western Powers to accept a Soviet demand for a Peace Treaty with Germany, the East German Government began constructing the Berlin Wall, closing the border between East and West Berlin.

Estimated Capability of the Control System

During hearings in July 1961 before the Joint Committee on Atomic Energy (U.S. Congress, 1961) the U.S. Congress was provided a clear view of the scientific community's assessment of the capability of the Geneva system. It was presented by Dr. Richard Latter, at that time chairman of the Ad Hoc group on Detection of Nuclear Detonations.[37] The group had been established by the Department of Defense to provide scientific advice to ARPA on project Vela. As explained by Latter:

> For atmospheric tests the Geneva system appears to have a good probability of detecting and identifying nuclear explosions as small as 1 kiloton carried out at altitudes up to 6 miles, of detecting but not identifying nuclear explosions at altitudes between 6 and 30 miles. The Geneva System has a good probability of detecting explosions set off in the deep oceans. Since underwater chemical explosions and some natural events produce signals similar to underwater nuclear expulsions, identification may sometimes require inspection at the site. In some large areas of the oceans, particularly in the Southern Hemisphere, the capability of the system is degraded as a result of the great distances between control posts.
>
> For underground tests, it is believed that the Geneva system has a good probability of detecting and locating—but not identifying —underground nuclear explosions above about 1/2 to 1 kiloton conducted under the same conditions as the *Rainier* explosion. For fully decoupled explosions in salt, locating requires a yield greater than 150 to 200 kilotons. For underground explosions above a magnitude 4.75, that is for explosions above about 20 kilotons conducted under the same conditions as the *Rainier* explosion, approximately 50 to 70 seismic events per year out of a total of 100 to 140 per year in the USSR will not be identified by the Geneva system. These numbers of unidentified events vary from year to year due to variations in the annual number of earthquakes. These same numbers apply to fully decoupled explosions of 6,000 kilotons in a salt medium. Unidentified seismic events require on-site inspection to determine what caused them. Since explosions of magnitude smaller than 4.75 are not subject to inspection according to the U.S. proposed treaty, these explosions, which can have yields from 20 kilotons under *Rainier* conditions

to thousands of kilotons fully decoupled, cannot be identified and cannot be controlled.

Latter went on to point out that "It is generally agreed that a cavity of at least 800 feet in diameter, sufficient to decouple at least 100 kilotons, is feasible!" Expert testimony on the feasibility of constructing such cavities had been presented to the same committee the previous year by L.F. Made of Phillips Petroleum (U.S. Congress, 1960). (Latter's mention of 6,000 kt in the paragraph above was obviously only to illustrate his point, not to claim such a large test was feasible.)The estimated number of earthquakes in the USSR (100-140) larger than magnitude 4.75 came from an independent analysis conducted by Rand Corporation scientists. They were similar to, but not identical to, estimates made by AFTAC and endorsed by the Berkner Panel.

The USSR Resumes Nuclear Testing

Then, during the night of 30 August 1961, the Soviet Union announced that it would resume nuclear testing. In a lengthy tirade against the Western powers (U.S. State Department, 1961), the Soviet government stated that "the United States and its allies are spinning the flywheel of their military machine ever faster, whipping up the arms race to an unprecedented extent..." Referring to the nuclear test ban negotiations, they accused the United States and Britain of "doing their utmost to prevent agreement." In fact, they deleted the unanimously adopted conclusions and recommendations of scientific experts, including their own experts—American and British—concerning the methods of identifying nuclear explosions... "Western powers," they said, proposed to "cover the territory of the Soviet Union with a network of espionage centers in the guise of control posts and teams." They pointed to "much ado" within the United States to develop a neutron bomb. "Such a bomb would kill every living thing but at the same time would not destroy material objects. Only aggressors dreaming of plunder" would mobilize scientists for developing such weapons, they said. The statement asserted that "the United States is standing on the threshold of carrying out underground explosions [seismic research and Plowshare explosions] and only waiting for a pretext to start them," they cited prior warning that "the Soviet Union will be forced to resume tests if France does not stop her experiments." They stated that "the Soviet Union has worked out designs for creating a series of super-powerful nuclear bombs equivalent to 20, 30, 50 and 100 million tons of TNT" and that they had the means to delivery them to any part of the globe.

The intemperate Soviet diatribe was followed by a test detonated the next morning. The Geneva Conference recessed on 9 September 1961, with no definite plans to reconvene. What followed in the next nine weeks was the largest nuclear weapons testing program the world had seen up to that time. There were tests at Novaya Zemlya, at Semipalatinsk, and at their Missile Test Range on rockets fired from Kapustin Yar (north of the Caspian Sea). (Please refer to map, Figure 11.1) There were up to three tests on some days. They fired the largest explosion of all time: 50 megatons (we estimated 58 Mt at the time); and analysis of its radioactive debris showed that the size could be doubled. The shot produced acoustic waves so strong that they were detectable by microbarographs—and long-period seismographs as well -- even after circling the globe several times. They fired two explosions in space, and one underwater. And unknown to seismologists outside of AFTAC, the USSR fired its first underground nuclear explosion on 11 October 1961. (Six seismic stations reported signals to the USC&GS. The location published by the C&GS was near the Semipalatinsk test site, but at a depth of 31 km, and thus the event was believed by most seismologists to be an earthquake).

It was a time at AFTAC that I well remember. Many of the tests could be detected seismically, and we were kept busy calculating and refining their locations. For the larger shots we could also estimate the height of burst using the yield measured acoustically and the time difference between the electromagnetic pulse and the seismic origin time (the time the bomb's shock wave hit the ground). We were also kept busy deploying mobile seismic stations into the field for recording explosions at the Nevada Test Site that would inevitably follow.

Shortly after the Soviet announcement, the White House had denounced the Soviet decision to resume testing; and on 3 September President Kennedy and Prime Minister Macmillan proposed to Premier Khrushchev that the three countries sign an immediate agreement not to conduct tests "in the atmosphere and produce radioactive fallout." The proposal was rejected. The U.S. nuclear response did not come until the first shot of Operation Nougat almost two weeks later on 15 September —a 2.6 kt underground test named *Antler*. The U.S. was simply not prepared to test and, anyway, President Kennedy reportedly thought it wise to let the international condemnation remain focused mainly on the Soviet Union for a while. It soon became clear that the Soviets, in secrecy, had made extensive preparations—many experts believed they began no later than the collapse of the Summit talks 15 months earlier. No wonder the Conference in Geneva had made no progress for so many months! The Soviets had had no intention of banning tests and in fact, "the talks had been deliberately misused as a screen for test preparation," as Ambassador Dean would later write. (Dean, 1966).

The Geneva Conference Resumes, and Collapses

Under the urging of the United Nations, the Geneva Conference resumed at the end of November, in spite of the breakdown of the moratorium. It began, as might be expected, with a proposal by the Soviets for an immediate ban on all tests in the atmosphere, underwater and in space. The Soviet proposal asserted that existing national systems were adequate to monitor such an agreement, and therefore no international monitoring system was needed. It also called for a moratorium on underground tests until a control system could be negotiated as an integral part of an agreement on "general and complete disarmament."

For the U.S., this was a flagrant attempt to secure for the USSR unilaterally the advantages gained by the recently completed, enormous Soviet testing program. Beyond this, history suggested that negotiations with the Soviets on underground test monitoring might never be completed. Furthermore, the U.S. would not again agree to an unverified suspension of tests, let alone general and complete disarmaments, after seeing how the USSR had used the previous moratorium to prepare secretly to test. The talks produced no results, and eventually, at the end of January 1962, the three-nation talks collapsed. Similar talks would resume in March, however, in a new forum: the Eighteen Nation Disarmament Conference.

Gnome and Early NTS Explosions

With the resumption of U.S. underground nuclear tests in the fall of 1961, badly needed seismic data flowed into the research program; results were prompt and dramatic. In early December *Fisher,* a 13 kt device buried 360 meters deep in alluvium, was detonated at NTS. It was followed one week later by *Gnome*, a 3 kt explosion in a thick layer of salt near Carlsbad, New Mexico. *Gnome* was the first nuclear explosion detonated for peaceful purposes as part of the Plowshare Program. Its objectives had been defined as: to explore the feasibility of converting nuclear energy into heat for electric power production; to investigate the recovery of useful radioisotopes; to make neutron measurements for general scientific knowledge; and to explore the design of nuclear devices specifically for peaceful purposes. More importantly from our viewpoint, it was the first U.S. underground explosion outside of the NTS, and it was in an entirely different material than previous underground explosions. We had deployed our mobile seismic stations carefully to learn as much as possible from this unique event. On reflection, I believe that *Gnome* turned out to be the most revealing for our science of all nuclear tests.

About half of our mobile seismic stations, (46 of them, thanks to criticisms by our Russian colleagues at the SRPAG meetings) were deployed along two lines, one running east-north-eastward from the *Gnome* site, and the second running north-westward from *Gnome* through and beyond NTS. Both *Fisher* and *Gnome* were well recorded by stations along the latter line. Travel times and the rate of change of amplitudes with distance were identical in either direction. But the amplitudes themselves astounded us. The smaller *Gnome* shot had produced far larger signals than *Fisher*. In fact, adjusted for the yield difference, seismic coupling in salt was 40 times greater than in dry alluvium! This was in the opposite direction from that predicted from physical reasoning, by such experts as Albert and Richard Latter and their colleagues. Apparently our understanding of seismic coupling had been, at best, primitive—if not based on unsound principles.[38]

But *Gnome* had other surprises in store for us. Both travel times and amplitudes of P waves differed greatly along the two main profiles. At distances beyond a few hundred km and continuing to about 2200 km travel times were invariably less to the east and greater to the northwest. Differences of up to 10-12 seconds were observed at the same distances; or as *Gnome* helped teach us to think about these observations, seismic velocities were high in shields and old platform regions of the continent, and low in tectonically active regions. Amplitudes were generally larger to the east than to the northwest, sometimes much larger. For example, at Jackson, Tennessee, 1459 km to the east, amplitudes of P were more than 100 times larger than at Mina, Nevada, 1465 km to the northwest (see Figure 11.2).

Forty-three seismic exploration teams had also been deployed to record signals from *Gnome*. These teams, whose primary task was to explore for oil and gas, had been organized under the aegis of the Society of Exploration Geophysicists as part of a voluntary project to assist in the underground explosion measurement program. *Gnome* was the first explosion since *Rainier*, more than four years earlier, to be announced in advance. Arrangements had been made for radio broadcast of the countdown to detonation, and pulsed tones at one second intervals beginning with the detonation were broadcast as well. With this timing information, exploration teams equipped with recorders that could record only several seconds of information were able to make successful recordings of P waves and measure accurate travel times. Most of the exploration teams operated in Texas and Oklahoma, at distances less than about 550 km from *Gnome*, and few were calibrated to provide amplitude data (of questionable value anyway, for reasons previously given) but the travel times were useful supplements to data from our primary mobile stations.

Variations in the speed and time of arrival of the P-wave were not unexpected. Velocities of Pn observed from historical explosions in different regions [see, for example, Byerly, (1940), Gutenberg (1952) and Jeffreys (1947)] had reported velocities that differed to some degree. More recent observations [see, for example, Tatel and Tuve (1955), Berg, et. al. (1960), and Diment, et. al. (1961)] had reported even larger discrepancies. (Berg, et. al., had observed such low speeds in Utah and Nevada that they doubted that the waves were actually Pn). But each of the referenced studies had sampled relatively small regions of the earth's crust and upper mantle.

The *Gnome* data, however, sampled large regions of the United States, and tied the discrepant observations into a comprehensible pattern. Velocities to the east of the Rocky Mountain Front were consistently high (negative travel-time residuals), while most velocities to the west of that great geological divide were slow (positive travel-time residuals). Berg's low velocities were consistent with those elsewhere in and below the young, tectonically active, provinces of the Western United States; Tatel and Tuve's high velocities were also consistent with other velocities in the ancient platform and shield provinces of the east.

These large discrepancies in travel times in different directions brought a serious problem into sharp focus for seismologists: location bias. Determining the location of a seismic event was (and still mostly is) based on the premise that seismic velocities are uniform in all directions, and that identical arrival times of P-waves means identical distances from the source. It was expected that slight differences might exist due to measurement error or lack of homogeneity of the earth's crust, etc., but it was generally believed that these effects would be random and hence tend to average out. Time differences observed from *Gnome* were not random; they were highly systematic in direction. The consequence for calculating an epicenter was that locations would also differ in a systematic way from the true location (bias). Had we not known *Gnome*'s location, but instead dispatched an inspection team to a location based on the usual seismic methods, the team might have searched a region tens of miles to the east of the *Gnome* site. Perhaps even worse in our nuclear detection context, we could have concluded that *Gnome* was an earthquake, since its calculated depth according to one estimate was more than 100 km below the surface. (The locations differed widely, depending on the selection of stations whose data were used in the calculations.)

At the time of *Gnome*, high speed digital computers were gradually becoming available to seismologists, and an early application was to compute locations of seismic events. One of the earliest reports on this use was by Bolt (1960), who had demonstrated a capability to use data from as many as 200 stations for this function. With data from *Gnome* establishing systematic, mappable regional velocity variation, Herrin and Taggart

(1962) were able to extend the earlier work to correct for these velocity variations. They were able to show an order of magnitude reduction in travel-time residuals (errors) from *Gnome*, but reported that using only data from stations more distant than 200 km, the calculated depth was still about 50 km too deep. Nevertheless their work was a significant advance —efforts are still underway today to map velocity variations in various regions to improve epicentral location accuracy.

Two months after *Gnome*, on 15 February 1962, *Hard Hat* (the renamed *Lollipop* explosion) took place. As was the case for *Gnome*, the P-wave amplitudes from the 5.7 kt explosion in granite were much larger than those from the higher yield *Fisher* explosion, again demonstrating that seismic coupling is greater in rigid, hard rock than in weaker tuff or alluvium. The *Hard Hat* explosion also generated relatively strong Love and Rayleigh waves; more about these later.

Underground Explosions Continue

The USSR fired its second underground explosion on 2 February 1962; with a magnitude of about 5.3 it appeared to be the largest underground explosive event to that time. It was recorded by 35 LRSM mobile stations as well as 9 stations of the WWSSN.

Relatively low yield underground testing continued at the Nevada Test Site during the winter and spring of 1962. Our seismic measurement program continued to monitor these tests and we were able to determine the dependence of amplitude on yield in both tuff and alluvium, as well as to measure the relative amplitude of a single shot in granite. We soon discovered that shots in the tuff of *Rainier* Mesa generated larger seismic waves than shots in tuff underlying Yucca Flats, and began to differentiate between "mesa tuff" and "valley tuff and alluvium" (see Figure 11.3). The role of gas-filled void space in these materials, which originated from volcanic ash, began to be recognized as a cause of "muffling" seismic signals.

What continued to elude us were direct measurements of the magnitude, m_b, for any U.S. shot; the shots were simply too small and too poorly coupled to produce adequate teleseismic signals. After the bitter disagreements of TWG-II, we could settle for nothing less than reliable teleseismic measurements of m_b to calibrate the test site.

The U.K. Array Program

Research on seismic detection was also underway in the U.K. among scientists at the Atomic Weapons Research Establishment (AWRE) near Aldermaston, England. Their work soon focused on seismometer arrays, using broad-band sensors of their own design. A main thrust of the program was to explore the concept of "velocity filtering," i.e., using an array to enhance (or to discriminate against) P-wave signals according to their speed and direction of arrival. To do this most effectively, the array must be comparable to, or larger than, the seismic signal wavelength, in contrast to the "Geneva array, " designed to be small compared to teleseismic signal wavelengths. However, to combine the signals coherently, the large array must have a way of correcting (delaying) the differing P-wave signals to a common time of arrival.

As previously noted, the concept of velocity filtering was known to the Berkner Panel, and work was underway in the U.S., most notably at Texas Instruments, Inc., to explore its capabilities. The thrust of the two programs differed, however. In the U.S., the initial objective was to attempt to find velocity characteristics of the seismic noise that would permit reduction of it beyond $n^{-0.5}$, the theoretical reduction for random noise at n detectors. The focus was also on defining the maximum capabilities of small arrays, as specified by the Conference of Experts. We would give the Soviets no further excuse to reject our research on narrow legalistic bases as they had done before! However, we would not ignore the possible benefits of larger arrays, and work had begun on such an array in Arizona.

As the U.K. program developed, largely under J.W. Burtill and F.E. Whiteway initially, AWRE deployed 6 to 19 seismometers spaced 1 to 2.5 km apart along two perpendicular lines. Data from each seismometer were recorded on one channel of a newly-acquired 21 channel tape recorder. To add the signals coherently, they constructed an "analog computer with facilities for applying time delays to the individual channels " (Carpenter, 1965). Using a method they referred to as the "correlation technique," the outputs of seismometers along each line were delayed and summed appropriately for the direction and distance of the known epicenter of an event to be studied. Then the two summed outputs, virtually identical signals, were multiplied together and smoothed over a short time interval.

By mid-1961 preliminary tests of this system had been conducted in England. As plans for the *Gnome* explosion became concrete, the AWRE group rushed to install a six element array at the site of the abandoned AEDS station at Pole Mountain, Wyoming. *Gnome* was successfully recorded, as were other tests at NTS and the Soviet underground explosion on 2 February 1962.

Word of this success reached high level officials of the U.K. government, who were soon in contact with the White House. As a consequence, senior scientists from AWRE were invited to present their results to U.S. scientists, which they promptly did. I cannot date the meeting precisely, but recall that it was held in early 1962 over a weekend in a conference room at the State Department, chaired by Dr. "Jerry" Wiesner, the President's Science Advisor. The U.S. and U.K. were considering changes in Test Ban Treaty negotiating policy, and the U.K. was pressing for abandonment of the requirement for seismic stations inside the USSR, hoping that their array research would convince us that such stations were no longer needed.

U.K. scientists presented examples that seemed to demonstrate phenomenal increases in signal-to-noise ratios (SNR). Uncritical acceptance of this probably prompted the enthusiasm of the senior U.K. officials. However, as I pointed out, their correlation technique was non-linear, and large increases in SNR on signals recorded at high SNR did not demonstrate that the technique would give similar results on signals recorded at low SNR. (To appreciate the underlying problem, consider what happens when two identical recorded signals and noise are multiplied together. At the risk of greatly oversimplifying, if the ratio of signal to noise is greater than 1, say 3/1, then the product is 9/1 and the SNR seems to be increased. But if less than 1, say 1/3, then the product is 1/9—and the detectability of the signal is worsened.)

However, a serious limitation of the large U.K. array, in its then existing state of development, was that it required independent knowledge of the epicentral coordinates of small events. Without such coordinates, the seismometer outputs could not be corrected for the differing arrival times as the signal swept across the array. Thus, it could not be applied to the critical task of *detecting* small events. The alternative strategy would be to monitor all possible epicentral locations by forming a large number of simultaneously processed outputs (or "beams," in the parlance of seismologists today). In the words of Carpenter (1965), "This, it was shown, could not be done without three stretch computers[39] running in parallel!"

U.S./U.K. negotiating policy was not changed as a result of the visit. The large U.K. arrays would certainly be useful for detailed analyses of signals, such as separating overlapping signals from different events, or highlighting different phases from a single event. They could be very valuable in establishing focal depth by reflected waves, for instance. But they were disadvantageous relative to the omnidirectional "Geneva stations" in the critical function of detecting and locating small events—the technology to support this function by large arrays was simply not available at that time.

U.S. Atmospheric Testing Begins

The Soviet tests of the previous Summer and Fall had caused concern in the U.S. that the Soviets might have gained the lead in a number of critical aspects of nuclear weapons design. In his address to the nation on 2 March

1962, President Kennedy said of the tests, "they also reflected a highly sophisticated technology, the trial of novel designs and techniques, and some substantial gains in weaponry." Preparations for atmospheric testing had been authorized early on, but the President withheld authorization to actually test until March of 1962. On 25 April 1962 atmospheric testing resumed in the Pacific with the 190 kt *Adobe* shot. The U.S. had finally begun a serious "catch-up" operation in response to the massive Soviet program. This time, however, rather than returning to test in the Marshall Islands, U.S. tests were conducted near Christmas Island. The island, a low-lying atoll in the Line Islands, near the equator (the Line) south of Hawaii, had been claimed by the United States in the mid-nineteenth century. However, it was administered as part of the Gilbert and Ellice Islands Colony by the U.K., which had conducted nuclear tests there in 1957 and 1958. At about the same time the U.S. began testing in the British colony, the United Kingdom began testing underground at the Nevada Test Site, a practice that continued at a low rate until the end of U.K. testing in 1991.

Atmospheric testing of devices as large as 7.7 Mt continued near Christmas Island until mid-July, with one exception: a 1.4 Mt test near Johnston Island. There was also a low yield deep underwater test, *Swordfish*, west of Baja California. Its yield has not been released, but it was large enough that we obtained recordings from 54 seismic stations (and we used the data as part of a study of location accuracy).

Although treaty negotiations were deadlocked over verification issues, advances were being made in the technologies underlying verification. Large atmospheric explosions in the Pacific were increasing the accuracy of travel time of seismic waves, and hence accuracy in location of seismic events. Smaller underground explosions were adding to knowledge of travel times, and improving knowledge of coupling, and hence of our ability to measure yields of such explosions. The AEDS retrofit program would add to our seismic capability, and soon play a role in our negotiating strategy.

TWELVE

Revised Seismicity and Network Analysis

Worsening relations between East and West, and lack of progress in the trilateral negotiations on a nuclear test ban prompted member states of the United Nations to become vocal critics of the major powers. The Soviet abrogation of the moratorium on testing in 1961 added the issue of radioactive fallout to the clamor. While the U.S. seismic research program benefited from the underground tests that were the initial response to the Soviet resumption, U.N. members remained critical of even this restrained form of testing.

As a result, an "Eighteen Nation Committee," consisting of five NATO members, five Warsaw Pact members, and eight more or less non-aligned members was established by the United Nations at the end of 1961. It was intended as the principal forum for international disarmament discussions. Its first meeting was held in Geneva in mid-March of 1962. Shortly thereafter a "Subcommittee on a Treaty for the Discontinuance of Nuclear Weapon Tests" was established. It consisted of the United States, the United Kingdom, and the USSR. The three nuclear powers were together again, and starting out with virtually the same positions that had already led to the collapse of tripartite negotiations in January of that year. The U.S. and U.K., however, dropped their proposal for a magnitude 4.75 threshold, returning to their original goal of a fully comprehensive treaty. They continued to require internationally manned control posts and on-site inspection. The USSR would accept no form of international verification, and accused the West of seeking unilateral advantages "to the detriment of the security of the other party" — a complete impasse once again.

The Eight Nations' Suggestions

In this unpromising atmosphere, the eight non-aligned nations[40] banded together, and on 16 April 1962 they proposed a set of "suggestions and ideas" for the nuclear powers to consider as a basis for new negotiations. Their three principal points were:

1) Establish a control system, "based and built upon already existing national networks of observation posts and institutions" or "if more appropriate," on existing observation posts supplemented by additional posts established by agreement. "Improvements could no doubt be made by furnishing posts with more advanced instrumentation."

2) Set up an international commission of "highly qualified scientists, possibly from non-aligned countries" to evaluate the data and report any explosion or suspicious event. Parties to the treaty would be required "to furnish the facts necessary to establish the nature of any suspicious and significant event," and could invite the commission to visit their territories "and/or the site of the unidentified event."

3) After consultation between the party and the commission on further measures, including "verification in loco," the commission would make its assessment of the event concerned.

The suggestions were vague and ambiguous; in response to questions from the West, one of the sponsors replied that it was up to the nuclear powers to interpret the suggestions. He simply hoped that they could become the basis for a new start to negotiations.

The Soviets promptly welcomed the ideas, but they were quick to interpret the proposals as permitting *only* national seismic stations, plus the international scientific commission. In their view, on-site inspection would take place only at the invitation of the country being questioned by the commission. The U.S. and U.K. promptly denied the adequacy of existing national stations to verify compliance with a test ban, and reasserted the need for mandatory on-site inspection of unidentified seismic events.

AFTAC External Network Study

Although not acceptable to the U.S., especially as interpreted by the USSR, the eight nations suggestions prompted ACDA to request that AFTAC conduct a study of what could be done to monitor the USSR and China based on facilities external to those countries.

Seismologists in AFTAC were quite familiar with many of the free world's seismic stations, and routinely used data from the more sensitive and reliably operated stations. Years earlier, AFTAC had made arrangements through the USC&GS for a selected set of stations to transmit prompt reports, either by paying the cost of their telegrams, or arranging for the data to be given to a U.S. embassy or consulate for prompt transmission. The data were used by the USC&GS for making prompt "Preliminary Determination of Epicenter" reports that were circulated internationally, as well as by AFTAC in surveillance of the USSR and China—chiefly to improve location accuracy.

For our analysis we selected a network of 21 of the more reliable stations surrounding Asia, including such familiar names (to seismologists) as Kiruna (Sweden), Hungry Horse (Montana), College (Alaska), Matsuhiro (Japan), Quetta (Pakistan), and Tamanrasset (Algeria). We either had measurements of noise levels for those stations, or for an "equivalent" station in a similar geological setting and distance from probable seismic noise sources. For our study we assumed explosions might take place at seven locations that we selected in the USSR, and two in China. We then went through hand calculations of the type previously described and found that the network had the capability to detect signals at four stations—the minimum for even a rough determination of location and depth—at about magnitude 4.5 within the USSR and China. This corresponds to about 6-7 kt under *Rainier* coupling conditions. Not bad, but not equivalent to capability estimated for the Experts' system.

Three Significant Large Underground Explosions

On 1 May 1962 France fired a large underground nuclear explosion near Haggar in what was then the French Sahara (now Algeria[41]). As was the case for the large Russian test in February, the French explosion was well recorded by seventeen stations of the Worldwide Standard Seismological Network (WWSSN)—forty had been installed by that time—as well as by thirty stations of our Long Range Seismic Measurements Program (LRSM) and we noticed a somewhat unexpected result: magnitudes measured from both shots at LRSM and WWSSN stations agreed quite well with magnitudes measured by the AEDS seismic network.

Since the Hardtack II explosions, the first underground nuclear explosion in Nevada large enough to be recorded at numerous teleseismic stations was *Aardvark*, a 40 kt shot on 12 May 1962. Unfortunately, it was conducted in "valley tuff," which reduced the seismic signals below those expected from a shot in *Rainier* Mesa. Nevertheless, P-wave signals were well recorded at 15 stations at distances beyond 16° and weaker signals, but still usable signals for estimating magnitude, at 9 others. The average magnitude based on all 24 teleseismic signals was 4.7; using only the 15 best signals it was slightly less.[42] Magnitude 4.7 was also consistent with magnitudes of *Logan* and *Blanca* after allowing for the coupling difference between tuff in Yucca Valley and in *Rainier* mesa. Of most importance, the *Aardvark* magnitude was reasonably well determined based exclusively on m_b measurements at teleseismic distances.

An m_b Scale for Short Periods

This soon led to what was considered at the time to be the first break-through from the Vela program. The circumstances were these. Earth-quake statistics for the world, including the USSR, had been compiled by Gutenberg and Richter in terms of magnitude—but magnitudes measured chiefly at relatively long periods. Many magnitudes were on the M_s scale, based on surface waves measured at periods near 20 seconds. Some more recent earthquakes also included m_b measurements based on P-waves, but many of these included measurements at periods up to several seconds long, derived from long period seismographs. Gutenberg had believed he understood the relationships among these scales, but our Hardtack II measurements had cast some doubt on this.

We, in AFTAC, had been measuring m_b from Soviet earthquakes for several years, using data from AEDS seismic stations that employed short period sensors to record P-waves typically at periods of 0.5-1.0 seconds. Our directly measured statistics on numbers of events in the USSR at a given magnitude did not agree with estimates made in the established way, i.e. by extrapolating downward from large historical events in the USSR using magnitudes as reported in Gutenberg and Richter's "Seismic-ity of the Earth." Neither did our measured magnitudes of large recent events in the USSR agree with magnitudes of those same events as cur-rently reported by Gutenberg and Richter. To be consistent with Guten-berg and Richter magnitudes, we would need to assume our AEDS mag-nitudes were biased downward by several tenths of a magnitude unit. If we "corrected" for this possible bias, our statistics on numbers of current earthquakes also agreed with the numbers extrapolated from data on his-torical large earthquakes.

There were sound physical reasons to suspect such a bias might exist. We had noticed that our short period instruments gave lower magnitude estimates than long period instruments at the same site, both at AEDS and LRSM stations. We knew that long period instruments were common in many of the world's conventional seismic stations that had provided most of the data used for historical earthquake magnitudes. A second factor was that our AEDS sites had been selected to be on massive, hard rock (granite where possible), to take advantage of the low noise at such sites. But possibly signal amplitudes would also be lower than at typical sta-tions of other organizations, which in many cases are installed on sedi-mentary rocks or alluvium.[43]

What the *Aardvark* data did when we received it in late June, was give us a direct link between yield and mb based on short period instruments at sites more nearly representative of worldwide sites (LRSM and WWSSN). And this data, in turn, could be linked to numbers of earthquakes inside the USSR as a function of magnitude measured at AEDS stations. This followed from the fact that the AEDS magnitudes for the large Russian and French nuclear tests, and from *Aardvark,* agreed with magnitudes measured by WWSSN and LRSM stations. In short, we could adopt a new scale, based on teleseismic P-waves measured on short period recordings exclusively, and use the new scale for both earthquake statistics from the AEDS, and for U.S. explosions of known yields. It did not matter that the new scale differed numerically by several tenths from magnitudes reported by Gutenberg and Richter. It told us that the numbers of earthquakes in the USSR could be expected to average about 40 per year larger than the proposed magnitude 4.75 threshold, rather than 100 or more as Dick Latter had testified during the preceding years (see Figure 12.1).

As Frank Press explained it to the Joint Committee on Atomic Energy of the U.S. Congress (1963) a few months later:

> Until very recently the seismicity of the USSR for earthquakes with magnitude less than 5 was very poorly known. In fact, it could only be estimated by an involved and awkward procedure using large earthquakes in the USSR, large and small earthquakes in California and New Zealand, and an uncertain relationship between the surface wave magnitude scale and the P-wave magnitude scale. My colleague at CalTech, Prof. Charles Richter, the co-inventor of the magnitude scale, had, on numerous occasions, expressed doubts to me about this procedure because these small magnitude events were out of the range of validity of the equation relating the two magnitude scales.
>
> The new procedure reported by Dr. Romney in arriving at the new figures for the number of earthquakes in the USSR is simple and direct. We can now count the number of earthquakes in the USSR directly. We can use recently published Soviet data to check these numbers. We measure the magnitudes of these earthquakes with the same P-wave magnitude scale used for explosions. Thus we arrive at a more precise estimate of the number of earthquakes in the USSR corresponding to explosions of a given yield. This number turns out to be smaller by a factor of 2 to 3 than had been estimated earlier.

We in AFTAC recognized the significance of the new findings. We were simultaneously completing the study previously mentioned of the capabilities of existing seismic stations outside of the Soviet Union to detect and identify Soviet nuclear explosions. Our new information obviously had a direct bearing on the number of earthquakes that might remain unidentified. Several of us worked over the last weekend of 30 June and 1 July to be certain of our conclusions.

AFTAC External Network Study (continued)

Existing conventional seismic stations had been found, in our study, not to achieve U.S. goals for detection of seismic events inside the USSR and China. But what if the AEDS were to constitute the external network? Now that we had yield calibrated to magnitudes measured by the AEDS, and Soviet earthquake statistics on the same scale, the troublesome question of magnitude bias no longer existed. And we were certain that the AEDS network of carefully placed stations, equipped with arrays similar to the "Geneva stations" at Wichita Mountains and elsewhere, would have greater capability than a set of conventional stations.

What we found was that the recently modernized, AEDS network—which by then had been expanded to include stations in Flin Flon, Canada, Sonseca, Spain, Chiang Mai, Thailand, Mindanao, Philippines, and Alice Springs, Australia—could detect seismic events very near magnitude 4.0 at four or more stations throughout Asia. Furthermore, if supplemented by data from five stations inside the USSR, at sites selected to improve both depth determination and direction of first motion, we estimated the threshold was magnitude 3.9, or slightly less—a comforting 1 kt to political ears and essentially the detection threshold that was the original goal of the Experts. But more importantly, the supplemented AEDS network was estimated to be able to detect signals at 10 stations for shots as large as the U.S. proposed threshold of magnitude 4.75. With 10 signals, most deep earthquakes could be eliminated, and identification of earthquakes by direction of first motion at the U.S. proposed magnitude 4.75 threshold would be reasonably effective. In addition, location errors would be reduced to the order of 10-15 km in regions where seismic velocities were relatively constant in different directions.

Our results were highly significant: even though the Eight Nations' proposal did not result in a network that, by itself, satisfied U.S. goals, the U.S. could achieve them on a secret level provided that the Eight Nations' network included five stations at carefully selected sites inside the USSR. ACDA should be pleased with the results of our study when presented at a meeting scheduled for the coming Tuesday. First, though, we reviewed the work with our Technical Director, Doyle Northrup, on Monday morning, the 2nd of July. We also went over our new findings on seismicity of the Soviet Union. Should these reduced estimates be integrated into our study for ACDA? Or should we follow our normal procedure of having our results reviewed by a scientific panel first? We could later send an updated revision of the study to ACDA in the latter case.

Northrup quickly recognized that our new method of estimating Soviet seismicity also implied that, with fewer earthquakes, and improved AEDS detection and identification capability, the U.S. would need fewer on-site inspections than it was proposing in Geneva to achieve a satisfactory level of assurance about Soviet compliance with the test ban treaty. He immediately phoned John T. McNaughton, General Counsel to the Secretary of Defense who frequently oversaw test ban matters on behalf of Secretary McNamara. Northrup briefly described our new conclusions and asked whether our paper should be reviewed at a higher level before presentation to the interagency group. McNaughton thought it best to proceed with the presentation, including the reduced earthquake numbers, lest we later be suspected of delaying or withholding information.

The AFTAC Study is Presented to ACDA

An interagency group had been invited by ACDA to hear our results in a top-secret "tank" (as such highly secure facilities were commonly called) at the Department of State on 3 July 1962. About 25 people attended, including high level representatives of the White House, AEC, DOD, State Department, ARPA, and ACDA. Adrian Fisher, Deputy Director of ACDA, chaired the meeting. It was a typical conference/briefing room of the time — a long table in a smoke-filled room with a vugraph projector, a screen, and a lectern at one end.

Our SECRET level report, "An Estimate of the Detection Capability of a Network of Standard Seismographic Stations" was briefed to the group. As previously mentioned, the detection level of the unclassified network of national stations was at about 6-7 kt under *Rainier* coupling conditions — substantially above the level the U.S. thought necessary. Nevertheless, we could achieve our 1 kt detection goal on a classified basis, with the AEDS supplemented by five stations at selected locations inside the USSR.

I no longer recall our best estimate of the number of unidentified events. The unclassified working papers I have been able to retain are insufficient to reproduce that result. U.S. negotiators had proposed a quota of 12-20 on-site inspections to investigate an expected 70 unidentified seismic events annually larger than magnitude 4.75. We were now fairly sure that only about 40 events larger than magnitude 4.75 would actually occur, and our new study indicted that most of these would be identified as earthquakes by first motion and there would be good indications of focal depth on any remaining.

Our briefing was followed by numerous questions, and to our surprise, a number of questions from members of the ACDA staff amounted to "How long have you been suppressing this information?" We were stunned! Our previous estimates had been based on the best available classical data on Soviet seismicity. Our methods had been reviewed on a number of occasions by scientific peers. A completely independent study recently had been conducted by scientists at the Rand Corporation, producing closely matching estimates and had been the basis for Dick Latter's recent testimony before Congress. Now we were being challenged as to why we hadn't produced our new and more favorable results earlier.

We patiently explained again the critical role of data from the three large explosions, and especially *Aardvark*, only weeks before. Our relationship between magnitude and yield had not been accepted by the Soviets at TWG-II. Now we had independent confirmation, and much more data that could not be easily challenged. As soon as we had the actual data necessary to make better estimates, we had done so, and corrected our previous estimates. I understood that procedure to be part of the normal scientific process—step-by-step improvements, as new information becomes available. Apparently others, trained more in political skills, did not understand the scientific process, and preferred to search for hidden motives instead.

Perhaps even worse in its consequences, several of the more senior officials expressed concern that the information might be leaked to the press. Imagine! Every person in that room was cleared for access to many of our country's highest secrets, but it was suspected that one or more of them might give our secret results to the press (and thus to the Soviets), to the embarrassment of our negotiators in Geneva who had been arguing for a larger number of on-site-inspections than now seemed necessary.

The information obviously should not be suppressed. It was highly relevant to the negotiation in Geneva; and therefore, just like the discouraging news from Hardtack II, this promising news must be shared with the other Conference participants. But perhaps it should be delayed until our new estimates could be reviewed and validated by other seismologists, and the U.S. could reformulate its negotiating strategy in Geneva. In the end it was agreed that the Department of Defense would release a general progress report on results coming from the entire Vela Program, including a watered-down version of our study. The task of arranging for scientific review and preparing such a press release fell to Dr. Jack Ruina, the Director of ARPA.

Ruina was an electrical engineer on leave of absence from an academic career to serve in several senior scientific positions within the Department of Defense. Although his responsibilities in ARPA were very broad, he had taken a keen interest in the nuclear test detection research in his agency, and personally represented the program in many key meetings. He was well prepared for the task.

The Defense Department Press Release

Somehow, Ruina managed to assemble a group of independent seismologists to review our work within a few days, and to select a set of recent significant results from the ongoing research program. The press release, dated July 7, 1962, explained the objectives of the Vela research, then cited a number of results. Progress had been made in detection methods by using detectors in deep wells, citing "initial research results" that indicated a five to ten times increase in sensitivity might be achieved, as was the possibility of establishing stations in areas previously believed to be too noisy for detection stations. It mentioned that further experimentation with seismometer arrays, using special filtering techniques, showed improvement in sensitivity "somewhat greater than had previously been considered possible." It also mentioned the possibility of combining deep well and array techniques. Seismometers had operated successfully on the ocean bottom. Seismic results from *Gnome* were mentioned, both the positive and the negative. The explosion program was improving our understanding of seismic wave propagation and enhancing identification capabilities including "a good possibility of considerably enhancing the identification of earthquakes by further improving depth-of-focus determinations." Concerning our new seismicity estimates, the release stated:

> Additional study of the comparative signal magnitudes of the various seismic waves produced by nuclear explosions and earthquakes has indicated that there may be substantially fewer earthquakes that produce signals equivalent to an underground nuclear explosion of given yield in tuff than had heretofore been expected.
>
> If this is confirmed, it means that there will be fewer earthquakes that might be mistaken for possible underground nuclear explosions of a given size.

Strictly speaking, our seismicity study applied only to the Soviet Union, but no matter—we soon had evidence that it applied more generally.

Although the press release was deliberately written in tentative terms, and contained numerous qualifications, there were several solid accomplishments underlying it. Two that I would single out in addition to improved knowledge of Soviet seismicity, were improvements in detection sensitivity by placing seismometers in boreholes, and improvements in depth-of-focus determinations.

Concerning the use of seismometers in boreholes, it had been shown by 1962 that noise in deep boreholes at some sites could be as small as 1/30 of the noise on the surface (see Figure 12.2). There was a small reduction in P-wave signal amplitudes as well, but the signal-to-noise improvement was more than a factor of ten. At other sites the noise reduction might be much less. The reasons for these site-to-site differences were imperfectly understood, but there was a growing awareness that at noisy sites, deep boreholes offer a method of achieving signal-to-noise ratios comparable to those at very quiet sites (see Douze, 1964).

Perhaps even more important, it was found that seismic noise generated by wind and other local sources—a problem that plagued many sites that were exceptionally quiet when calm weather prevailed—could be almost completely eliminated by placing the sensors at modest depths of 30-50 meters. Herrin (1982) would later write:

> The seismometer-amplifier combinations discussed in this paper are the 'state-of-the-art' systems now in use or in development for use in treaty verification research. Only systems that can be placed in boreholes with casings of 7 inches or less in diameter fall into this category, because the borehole environment has been found to provide the optimum stability and lowest ambient background noise at any given site.

On depth determination, analysts in AFTAC, primarily Billy Brooks and his co-workers, had demonstrated that the depths of numerous shallow earthquakes in the USSR could be determined with high confidence. The method used was based on detecting reflected waves from the earth's surface above the earthquake's focus. Such use of reflected waves had been known to seismologists for decades, but was generally deemed ineffective at depths less than 60-70 km, the depth range generally taken to define "shallow earthquakes." Brooks and his group showed that depths of many USSR events as shallow as 15 km could be determined.

When such reflected waves can be observed, the time interval between the direct, initially downward-going P-wave and the initially upward-going P-wave reflected from the surface above is a measure of the depth of focus. This interval also changes with distance to the detecting stations. The reflected wave, however, arrives in the continuing motion of the P-coda (motion in the tail of the initial P-pulse). In conventional earthquake location work, seismologists had to rely on telegraphic reports of time and amplitude of P waves and so-called "secondary phases" based on signals large enough to be considered significant by the individual station operators.

The AFTAC group, however, had available the actual seismic recordings from a worldwide network of AEDS stations, transmitted to them almost instantaneously by the Zipagram system. They also had the capability of aligning the waveforms from different stations in time and distance and projecting them so that they were in near-juxtaposition. This allowed the analyst to see patterns of recorded signal arrivals that might not have been reported from the separate stations alone (see Figure 12.3). By formulating strict requirements on the signal-to-noise ratios for both P and the reflected waves (called pP and sP), as well as agreement with the calculated change with distance in the interval between the two pulses ("moveout," in seismological terms), the AFTAC group had found good evidence that about half of all earthquakes in and near the USSR occurred at depths of 15 km or greater.

Depth determination of shallow events had been considered by TWG-II, but had not been considered positive enough to qualify as a positive identification criterion—like the direction of first motion. Instead, the technique was categorized as a "diagnostic aid." Now it was taking on enhanced status as a positive identifier. Its significance for identification of Soviet earthquakes can be inferred from Figure 12.1, which shows that about half of all "shallow" earthquakes in and near the U.S.S.R. may occur at depths greater than 1.5 kilometers.

Looking back on 1962, the year may be seen as one in which progress was being made on a number of fronts. The most important from the point of view of this book were improvement in relating explosion signals to those from potential false alarms—earthquakes—and in identifying potential false alarms from more positive measures of their depth. But gaining better understanding of noise through array studies and deep well measurements was also important as was the increasing knowledge about the dependence of explosion signal strength on rock type. Serious problems remained for seismology of course, and treaty negotiations were stagnant, but I remember it as a time of optimistic confidence in the direction of our work, and the work of our colleagues.

THIRTEEN

Continuing Research and a Limited Treaty

Reaction to the Press Release

The Vela press release of 7 July 1962 attracted much attention nationally and internationally. Members of the Eighteen Nations Disarmament Conference (ENDC) took it as a signal that the U.S. was about to change its position in the treaty negotiations. U.S. and foreign press corps members scrambled for further clarification. The response of U.S. officials was somewhat confusing and uncoordinated, underscoring the lack of clarity in the U.S. position. Upon his arrival at the Geneva airport on 14 July 1962 for a new round of negotiations, U.S. ambassador Arthur Dean was accosted by reporters, and later quoted as saying that, under conditions described only in general terms, "perhaps we could do without them [control posts]." Emphasis to this incident was given by the almost coincidental publication of a lengthy article in the New York Times the following day. John Finney, a reporter for the Times had prepared a thorough report on the status of the treaty negotiations as well as on the Vela program and the recent DOD press release. Featured in the editorial section of the Sunday Times, the article was widely read nationally and internationally.

Secretary of State Rusk and ACDA's Deputy Director, Adrian Fisher promptly cabled Ambassador Dean that the U.S. had made no decision yet on whether or how to change its negotiating policy and they also issued a press release stating:

> The recent results of the Vela program, although promising, are of a preliminary nature and need to be fully evaluated before they can become the basis of any modifications in the United States test-ban proposals. These findings do not demonstrate the possibility

of doing away with control posts and on-site inspections to determine the precise nature of suspicious events.

The United States is evaluating and seeking further substantiation of these findings and will make in the near future whatever modifications in its present position as seem possible.

But the damage was done and when the ENDC resumed its meeting on 16 July, after a recess of several weeks, the Soviet representative seized on the Vela findings, claiming they supported the Soviet positions that neither control posts nor on-site-inspection were needed. The representative of India, claiming neutrality but clearly aligned with the Soviet Union on test ban matters, echoed these claims. In the absence of new U.S. policy, Ambassador Dean could only respond weakly, along the lines of the Rusk press release. He did, however, offer to bring scientific experts to Geneva to explain the full implications of the new findings.

The White House is Puzzled

President Kennedy was subsequently reported to be extremely irritated that the findings were being discussed publicly—and by opponents in Geneva—before the U.S. had had time to revise its negotiating policy. One consequence was that Doyle Northrup and I were summoned to the White House for questioning by Carl Kaysen, Deputy Special Assistant for National Security Affairs. Kaysen, an economist from Harvard, wanted to know about the circumstances and timing of our development of the new estimates of Soviet seismicity. We were unaware of the President's irritation, but it became clear within the first several minutes that Kaysen was investigating what he considered to be a very serious matter.

We patiently explained the earlier method for estimating the numbers of earthquakes and their relation to yields of underground nuclear explosions. And how we had found reasons, first to question some of the underlying "classical" relationships, and then to replace them as soon as factual data became available from recent large underground explosions —most importantly the *Aardvark* explosion.

In a memorandum to the President on 20 July (Mabon et. al., 1995) Kaysen responded to the President's question on "how long the Department of Defense had known the new figures on which they made their announcement...?" To this direct question he responded "There is no simple answer to this question in terms of time." Although he incorrectly attributed the discovery of the new data to the study requested by ACDA, Kaysen explained the uncertainties in the earlier estimates of numbers, and the new data that caused them to be questioned, reasonably well. Unfortunately, one of his major points was:

The estimates of the frequency of small shocks in the Soviet Union which we presented to the Geneva Experts Conference were challenged by Soviet scientists. Their figures differed from our figures by roughly the amount of the correction factor which we have now accepted. Thus our new figures are about the same as those the Soviets offered in 1958. We still do not understand whether this is a coincidence.

Where this misinformation on earthquake frequency came from remains a mystery. Although Pasechnik had described the seismic regions of the USSR, there had been no such quantitative discussion of the annual numbers of small shocks in the Soviet Union—only of the world. Both sides had accepted the procedure of extrapolating from the annual numbers of large shocks as well as the method of extrapolation, to estimate numbers of small shocks. Both sides accepted Gutenberg-Richter as the authorities on magnitude of large shocks. There had been minor disagreement on the number of world-wide small shocks because the estimates differed slightly between their 1954 publication (which the Soviets used) and Gutenberg's more recent 1956 publication (which we used). The misinformation may seem a small matter, perhaps, but the message it conveyed to the President was that the Russians had given us accurate information four years ago and that we had refused to accept it, or perhaps even suppressed it.

Kaysen went on to draw a "moral," speculative, he admitted, that perhaps if AFTAC data had been more regularly reviewed by outside scientists, our conclusions might have come to light earlier. How this could have happened prior to *Aardvark*, he didn't explain. Of course, we knew nothing of this letter at the time, or for many years afterward. Its effect, however, became apparent to us almost immediately.

The following day a TOP SECRET National Security Action Memorandum No. 174, addressed to the Secretaries of State and Defense and to the Director of Central Intelligence was issued by the White House. The memorandum began with:

The recent reports of new data on our ability to detect small underground shots from a distance came to our attention in circumstances that leave me puzzled. If Air Force officials had this data earlier and neither understood it themselves or consulted at an appropriately early time with the Disarmament Agency or the CIA, we need to improve our procedures.

It continued with "I do not understand why, in any event, the new interpretations were first presented to a large group of officials so casually and so near the date of the resumed negotiations." The three agency heads were requested to "consider how matters can be reorganized to avoid a repetition of this unfortunate event." Polite words, perhaps, but the meaning was clear: someone had badly fouled up, and I was at least one of the suspected culprits!

Fortunately, Adam Yarmolinsky was assigned to investigate. He was an experienced lawyer, serving as Special Assistant to the Secretary of Defense. When Northrup and I met with him, we were soon at ease. He quickly understood the elements of our recent analyses, and of the circumstances of our meeting with ACDA. The letter to the President that he prepared for Secretary McNamara stated flatly, "The information could not, in my judgment, have been developed at an earlier date." He was equally direct about responsibility for dissemination of the information at the meeting:

> Dr. Romney consulted his superiors about the form and content of his report, including the new evidence that the numbers of earthquakes in the Soviet Union might have been previously overestimated. They instructed him to include this information, since they thought it was relevant and timely.

McNamara's letter was signed 28 July 1962.

The Earth is Round

Three days later Doyle Northrup and I were asked by John McNaughton to brief Secretary McNamara on seismic detection of underground nuclear explosions. We were invited to sit with the Secretary around a moderate-sized conference table in his office. He explained that he had been corrected by an official of the State Department, on some matters related to our new information, but since the research was being done within his department, he felt he should become as knowledgeable as someone from State. He was cordial, clearly interested, and we were quickly at ease in his presence. The briefing was to be informal, and he would interrupt when he needed clarification.

After a brief introduction by Northrup, I told McNamara that I had brought a fair amount of tutorial information with me, and could begin with "the earth is round...", or at a higher technical level depending upon his wishes. He responded "Start with 'the earth is round'..." So I did.

Shortly afterward we were briefly interrupted by John McCone, at that time Director of Central Intelligence. After some short, private transaction in another room of McNamara's office suite, McCone joined us at the table, and stayed until a lunch break at noon. We continued with McNamara after lunch, and throughout the afternoon, and possibly during the next morning as well.

Throughout the briefing/tutorial, McNamara took copious, careful notes, and was soon asking astute, perceptive questions. At the conclusion of the briefing, John McNaughton asked that we prepare a written version for the Secretary's records. The briefing had been quite free form, responding to numerous questions and explaining numerous graphics and tables. It seemed impossible to recreate the lengthy briefing in any reasonably accurate way. The solution came to me the following day: borrow McNamara's notes! We contacted John McNaughton, and found that not only was it possible, but he would have a courier bring the notes to us. The secret briefing, based on the Secretary's notes, and co-authored with Doyle Northrup, Wayne Helterbran and Billy Brooks, was sent to the Secretary of Defense on 12 August 1962. It gave a fairly comprehensive view of the state of knowledge at the time on detection and identification of underground nuclear tests.

Seismicity Confirmation

By mid-July we had received preliminary seismic results from the second relatively large underground explosion at NTS. *Haymaker*, a 67 kt shot, had been fired in alluvium in Yucca Flats. Although even less well coupled seismically than the *Aardvark* shot, it was recorded by 14 stations of our Long Range Seismic Measurements program at teleseismic distances, and by five of the World Wide Standard Seismic Network. The average magnitude was about the same as that of *Aardvark*, and again supported our conclusions about the relationship of magnitude to yield, when allowances were made for differences among rock types. This, in turn, added to our confidence in our determination of the numbers of earthquakes in the USSR equivalent to a given yield.

Even as the policy makers in Washington struggled with how to integrate the new information into treaty negotiations , additional supporting evidence for the new estimates of seismicity of the USSR was received late in July. It came in the form of a translation of "Zemletrasemiia V USSR," published by the Academy of Science of the USSR in 1961. It contained tabulations of numbers of earthquakes in each of seven regions, categorized by magnitude, for the years 1947 through 1956. In regions where the Soviet seismic station network was relatively dense, the average annual numbers matched the AFTAC numbers as a function of magnitude. The Soviet magnitudes were based on surface waves, and were most nearly equivalent to M_s. Our magnitudes were in m_b units, but we had evidence by then that there was little difference between the two scales in the relevant range of magnitudes, i.e., near magnitude 5. No exact comparison was possible, but the agreement increased confidence that our estimates were good.

In the Arctic and Kamchatka-Kuriles regions, where the Soviet network was sparse, (and where stations had only Kirnos "SVK, SGK" seismographs—not suitable for detecting low magnitude events—the Soviet numbers matched the AFTAC numbers for events of magnitude 6 and larger, but at magnitude 4.5 the AFTAC numbers were more than five times greater. There was only one rational explanation for this discrepancy: the U.S. AEDS, even though using stations located outside of the Soviet Union, had a lower detection threshold than the Soviet network for these far eastern regions! Perhaps a surprising conclusion, but it was a clear demonstration of the value of searching for very quiet sites for the AEDS stations, and equipping them with arrays of short-period sensors specifically designed to detect small, distant, events.

A New U.S./U.K. Treaty Proposal

Following the Defense Department's press release on results of the Vela research, there had been intense activity within the U.S. Government to reformulate our negotiating strategy in Geneva. By the end of July the Committee of Principals, meeting with President Kennedy, had settled on a new set of proposals for monitoring underground tests. Briefly, these were:

- an international commission to oversee development and operation of the network, and to evaluate the data
- a network of nationally manned, but internationally supervised seismic stations would be constructed "involving a substantially smaller number of stations" than the Experts had proposed, and the U.S. had endorsed in April,
- existing stations, as approved by the commission
- new stations built and operated by the commission where requested by a host country
- obligatory on-site inspection of unidentified events (but a smaller number of inspections than the 12-20 the U.S. had called for in its previous proposal).

U.S. Ambassador Dean presented these proposals to the Eighteen Nations Subcommittee on 9 August, with introductory words linking them to the substantial improvements in long range detection capability from the Vela program. The new U.S. policy assumed a network of 25 control stations, nationally manned but under international supervision. In the U.S. view, five of these control stations should be at selected locations inside the USSR. Although the latter point was not spelled out at this time, Soviet Representative Zorin promptly rejected the proposals, contending that the Vela findings merely confirmed that national detection means are quite adequate, and that mandatory inspection was unacceptable.

Ambassador Dean had earlier offered to bring technical experts to Geneva to help explain the U.S. position to representatives of the Eighteen Nations. We arrived in Geneva on 11 August 1962, and began meeting informally with representatives of those countries. The U.S. group was headed by "Jerry" Wiesner, the President's science advisor. Those I remember were Dr. Franklin A. Long, Assistant Director for Science and Technology, ACDA; Dr. Jack Ruina, Director, ARPA; Dr. Roland Herbst, Lawrence Livermore Laboratory; and me. The meetings took place in the U.S. consulate, and typically began with an overview of nuclear test detection methods and capabilities, followed by questions and answers. Overviews were presented by Frank Long or Jerry Wiesner, and seemed designed to emphasize the relative ease of detecting and identifying explosions in the oceans and atmosphere, and the difficulty of identifying underground explosions. I sensed a shift in U.S. policy being planned: yet another proposal to ban only atmospheric, underwater, and space tests.

Representatives of the eight "non-aligned" countries had been chosen primarily for skills and knowledge unrelated to science, understandably. Several were also rather reserved. It was difficult for us to know whether our discussion and explanations were understood but they seemed to appreciate our efforts—with the exception of Indian Ambassador Lall, whose comments and attitudes projected a view that science and scientists were only obstacles to a nuclear test ban.

Our group stayed in Geneva several days, then returned to the U.S. A few days later, on 27 August, the U.S. and U.K. jointly tabled, not just the atmospheric/underwater treaty that I had sensed coming, but two treaties! The most desirable, President Kennedy and Prime Minister Macmillan stressed in a joint statement the same day, was the comprehensive treaty. But, as the President said two days later, "If we can't get that because of the Soviet Union's reluctance to permit us to have an effective inspection system, then we would like to get the second [atmospheric/underwater treaty] because that would have an effect in the arms race and it would also have an effect, of course, on the problem of radiation."

Provisions of the draft comprehensive treaty called for international verification measures essentially as described by the U.S. in the beginning of August. The partial test ban treaty required no international verification measures. It banned tests underwater and in the atmosphere, and prohibited the deposit of radioactive debris from underground tests beyond the territorial boundaries of the testing country. Both treaties were promptly rejected by the Soviet representative. The conference recessed on 8 September 1962.

The Cuban Missile Crisis

In mid-October of 1962 U-2 surveillance flights over Cuba brought back photographs of almost-completed launch sites for Soviet missiles, followed within days by photographs of the missiles themselves. On 22 October President Kennedy addressed the nation and the world, describing the gravity of the situation, and the U.S. response. He announced that Cuba would be blockaded at sea, and that any ship containing offensive weapons would be turned back. U.S. policy, he said, was to "regard any nuclear missile launched from Cuba against any nation in the Western Hemisphere as an attack by the Soviet Union on the United States, requiring full retaliatory response upon the Soviet Union." The United States and the Soviet Union were on the brink of war—perhaps leading to nuclear war.

As the reader knows, faced with a potential holocaust, both sides showed restraint. The missiles were withdrawn and the crisis was successfully resolved. But the leadership on both sides was apparently profoundly impacted by the dangerous confrontation. The incident led to a more active exchange of personal correspondence between Kennedy and Khrushchev, and has been cited as a significant factor leading to the signing of a test ban treaty the following summer.

When the Eighteen Nations Disarmament Committee reconvened in November, there was little change in the positions of either the U.S./U.K. side or that of the USSR. In a mid-December letter to President Kennedy, drawing upon a misguided "Pugwash statement on test detection,"[44] Premier Khrushchev indicated a willingness to place three automatic seismic stations on territories of each of the three nuclear powers. He proposed specific sites for such devices in the Soviet Union, none of them within seismic zones. Known as "black boxes" at the time, these were, conceptually, a vertical seismometer and recorder sealed in a container to be handed over to Soviet seismologists, who would place them on a pier at the designated location. Khrushchev also accepted the principle of on-site inspection, although he offered only 2-3 per year on Soviet territory. About a week later Kennedy's response pointed out the inadequacies of the "black boxes," reassured Khrushchev that we would accept stringent measures to ensure that inspection could not be used for espionage, and reiterated the Western position that 8-10 inspections were needed annually.

Some Seismological Developments

In testimony in March, 1963, before the U.S. Congressional Joint Committee on Atomic Energy, Dr. Jack F. Ruina reported that funding for the underground detection portion of Project Vela had grown to $41.1 million during 1963. More than half was directed toward preparing for and monitoring U.S. underground nuclear explosions. This investment was paying off in terms of understanding the generation and propagation of seismic waves from explosions, as has been reported earlier in this book.

Other parts of the Vela Program had also received healthy funding, and the results of the research were advancing the science of seismology. By early 1963 about 70 stations of the Worldwide Standard Seismographic Network (WWSSN) were operational. This new source of calibrated seismic data made it possible for the U.S. Coast and Geodetic Survey to improve the quality of its service to the international seismological community. On February 3, 1963 Captain Robert A. Earle, Chief of the Geophysics Division of the USC & GS, had announced , in a letter to seismograph station directors of the world, "We are about to begin computing body wave magnitudes, m_b as defined by Gutenberg and Richter, for as many earthquakes as possible for publication in the Preliminary Determination of Epicenter cards." He proposed to base the magnitudes on P-waves only, as recorded on short period vertical seismometers of the WWSSN. The letter was accompanied by detailed instructions on how to measure the amplitude and period of a P wavetrain, as well as the format for telegraphic reports to the USC & GS. The letter marked the beginning of modern attempts to catalog earthquakes world-wide on a consistent, well calibrated magnitude scale. The program actually went into effect four months later.

Also, as reported at the referenced Congressional hearings, the Geneva stations network had expanded beyond the Wichita Mountains Seismological Observatory. By the end of 1962 we had the "Cumberland Plateau Seismological Observatory" beginning operations in Tennessee, the "Blue Mountains Seismological Observatory" operating in Oregon, the "Uintah Basin Seismological Observatory" operating in Utah, and the "Tonto Forest Seismological Observatory" operating in Arizona. Each observatory had an array geometry and spacing that differed from the others, and the Tonto Forest Observatory eventually had an array of more than 50 sensors.

In my congressional testimony on 6 March, I mentioned that I had previously reported that the Wichita Mountains seismological observatory (WMSO) at Fort Sill, Oklahoma, had demonstrated the capability to detect 90 percent of all magnitude 4.0 seismic events at distances of 2000 miles and more. Since that time, two other stations at similarly carefully selected quiet sites (Baker, Oregon, and Vernal, Utah, had also demonstrated similar capability. Noise levels at a third such site (near Payson, Arizona) gave promise of the same capability when the station became fully operational. For the fifth station of our simulated Geneva network, no such quiet site had been found within the region that satisfied network spacing requirements. The station near McMinnville, Tennessee, the Columbia Plateau seismological observatory (CPSO), nevertheless had a threshold near magnitude 4.5.

The McMinnville station was also the U.S. test bed for our initial efforts at "beam forming." Like the U.K. arrays, data processing was based on analog "delay-and-sum" technology; unlike the U.K. arrays, it was designed to search for signals at a number of azimuths and distances simultaneously. Design and development was conducted by scientists at Texas Instruments beginning in 1960. The prototype Multiple Array Processor was constructed in 1961, and tested on tape-recorded data from WMSO in the spring of 1962. When installed at CPSO, it formed 18 delay-and-sum beams simultaneously to enhance signals arriving from preselected directions and with specified velocities, and pre-filtered to optimize the signal-to-noise ratio. In effect, it "flagged" the time and direction of incoming signals for further analysis. Since it was omnidirectional in response, and could be applied to data from larger arrays, it was the first practical array processor for continuous on-line nuclear test surveillance. An excellent summary of the device has been given by Robertson (1992).

Unattended seismic stations were described to the Joint Committee by Jack Hamilton. He outlined the early thinking of the Berkner Panel, which had considered large numbers of simple stations, perhaps single vertical component sensors spaced at intervals of about 250 km, transmitting real-time continuous data to control posts. As the concept was extended to the "black boxes" proposed at the Pugwash Conference of 1962, the device would need to record and store seismic data for up to three months. Given the technology of the time, this would tax the available data storage capacity of state-of-the-art recorders, even if only a single component of motion were to be recorded. If the objective was simply to add to location capability, and to record the direction of first motion, and if large numbers of stations could be deployed, relatively simple "black boxes" might suffice, although the data would generally not be available in time to influence critical decisions on on-site inspection.

However, viewed against any concept that might be acceptable to the Soviet Union, the unattended stations would be few in number and therefore far between. In this case, to be effective in aiding national seismic stations to identify seismic events, they would need all of the instruments and capabilities of a first class seismic station. Hamilton described a first attempt at designing an adequate unmanned observatory producing

3-component short-period and 3-component long-period seismic data- a far cry from the "black box" concept. I note, in passing, that this concept, described by Hamilton, with improved security, sophisticated data authentication, and satellite telemetry was subsequently adopted by the United States during treaty negotiations in the late 1970's.

Seismic Discrimination Research

The "Geneva" observatories, the WWSSN, and the stations of the Long Range Seismic Measurement Program (LRSM) were all rich sources of seismic data from both earthquakes and explosions. These data became the basic resource in the search for discriminants between the two types of sources of seismic waves. The results were mixed.

Two examples are cited here to illustrate the kinds of research underway to develop discriminants based on long-period surface waves. The first example, by Brune and Pomeroy (1963), addressed (among other topics) the unexpected generation of Love waves (surface waves having transverse horizontal motion) from underground explosions, first noted from *Logan* and *Blanca*. These waves were even more evident from the 5 kt *Hardhat* explosion, which was detonated in granite. Asymmetries in the aziminthal pattern of Rayleigh wave amplitudes were also observed. Brune and Pomeroy's general conclusion for explosions at NTS was that the source mechanism for most shots in tuff and alluvium was consistent with the simplest explosion model: a sudden force outward in all directions. However, for other explosions at NTS, "more complicated forces also act at the source." For *Hardhat* the "pattern would be expected from an earthquake source, but it could not be generated by a simple explosion force." They go on to report that, "On the basis of the surface wave radiation pattern, the *Hardhat event would certainly be identified as an earthquake" in the absence of other, non-seismic data* [italics added]. They speculate that the cause may have been motion along cracks, or more probably, by tectonic strain release (strain energy stored in the rock, the cause of earthquakes). Once again, a predicted explosion "signature" based on simple models of explosion sources may work in some cases, but criteria based on such models may result in complete misidentification of an explosion under other circumstances.

The second example, by Brune, et. al. (1963), studied the amplitudes of long period Rayleigh waves relative to short period P-waves from earthquakes and explosions. It was thought that earthquakes might excite relatively larger low frequency waves than explosions. It was argued that this would be a consequence of the greater dimensions of the source region for earthquakes, viewed as ruptures along faults. Brune et. al. adopted the Richter magnitude, M_L, as the parameter they would use to characterize the amplitude of short-period waves. For the longer periods they measured the area within the "envelope" of the Rayleigh wavetrain. An envelope was defined by a procedure that took into account both the amplitude and the duration of the wavetrain; they called this parameter "AR." When values of M_L and AR were compared for a given event, AR tended to be larger for earthquakes than for explosions of equivalent M_L value.

Shortly after publication of these results, seismologists in AFTAC and elsewhere found that using m_b to characterize short period waves, and M_S to characterize long period waves, greatly simplified the measurements and was equally effective (see Figure 13.1). At the higher magnitudes, the population of earthquake measurements may be seen to be well separated from explosion measurements. Projected to lower magnitudes the populations would appear to mingle. In practice, a decision line is drawn parallel to the trend-line of the explosion data set to eliminate as many earthquakes as possible without misidentifying more than a selected percentage of explosions. In the example shown (Figure 13.1), the upper dashed line would correctly identify all Eurasian earthquakes larger than about $M_s = 4.5$ (below which the sparsity of data points indicates that most earthquakes have not been detected), at the expense of misidentifying 5% of the explosions. (The three explosion measurements inside closed brackets were ignored in the analysis: more on that later.) The criterion, often described as the "$M_s:m_b$ criterion," remains in use today.

The Vela Seismological Center

In early 1963, Dr. Jack Ruina approached AFTAC with a request that we establish a unit within our organization dedicated to management and technical direction of Vela projects. Under this concept, the unit would have offices separated from the highly secure AFTAC headquarters, which would facilitate visits by scientists from the university community, many of whom had no security clearance, as well as visits by foreign scientists. It would also allow the assigned AFTAC personnel to devote full time to Vela research. We welcomed this proposal, believing we could benefit from more contact with unclassified work, and without the distraction of other, sometimes conflicting, duties. Accordingly, we responded with a "Concept of Operation for an AFTAC/Vela Seismological Center" in April 1963. The Vela Seismological Center thus came into being, and operated in Alexandria, Virginia for the next decade, as a principal focus for Vela research and development.

Complexity

According to Carpenter (1965) the concept of P-wave complexity originated from "correlation" analysis of signals from the large Soviet underground test of February 1962 and the large French test of May 1962. The so-called correlation signals for these events, as defined in Chapter 11, were characterized by high correlation during the first few seconds after the onset of P, followed by much smaller correlation. Earthquakes, on the other hand, often produced correlation traces that persisted with significant correlation for thirty seconds or more after the onset of P.

It seemed plausible to the U.K. scientists to attribute this difference to factors such as differences in the radiation patterns of the two kinds of events. Explosions were believed to radiate P-waves uniformly in all directions, producing a strong, short pulse at all recording stations. Its coda (or tail) consisted of relatively weak scattered waves. Earthquakes, on the other hand, radiate compressional P-waves and rarefactional P-waves in four alternate quadrants, and near the boundaries between quadrants there should be regions where little or no energy is radiated. In such regions the initial P would be weak, and the scattered energy in the coda would be of comparable amplitude. The short, sharp pulse from an explosion was described as "simple," whereas the more sustained signal from earthquakes as recorded between quadrants were described as "complex."

Thus "complexity" came to be defined as the area under the correlation signal between 5 and 35 seconds after the onset of P, divided by the area between 0 and 5 seconds (see Figure 13.2). Some earthquakes were also found to produce simple P waves, but it was argued that only earthquakes could produce complex P waves.

Complexity was enthusiastically advocated by seismologists in the U.K. beginning about mid-1962. As data from large U.S. underground nuclear explosions began to become available, the method was tested on these explosions, as well as the Soviet and French test. U.S. explosions also seemed to strengthen the concept. When transmitted up the technical chain of command to U.K. political levels, and from there to U.S. political levels in the Spring of 1963, it was passed down to U.S. seismologists as a "breakthrough"—a positive identifier of both earthquakes and explosions.

The criterion became the subject of one of the early studies undertaken by the newly formed Vela Seismological Center (VSC). Extending the U.K. analysis to include far more signals, we found, as reported in Technical Note VSC-1, (VSC-1, 1963):

> Explosion signals are not exclusively simple, however, nor are earthquake signals exclusively complex...Recordings of hundreds

of explosion signals and thousands of earthquake signals show that waveform envelopes from both classes of events are distributed from simple to complex. The distribution curves seem to be different, however, with the peak of the explosion distribution curve tending to be toward simpler waveforms.

The analysis of the pulse form of P would thus appear to fall in the general category of "diagnostic aids" to the interpretation of seismic events, as used by Technical Working Group 2 in 1959. In the conclusions of TWG-2, "criteria" were based on methods of analysis (such as first motion) that relate directly to differences in source characteristics, while "diagnostic aids" were based on methods less directly related to the nature of the source. The analysis of the P wave envelope by itself neither identifies earthquakes nor explosions.

Subsequent studies at VSC (VSC-5, 1964) developed strong evidence that much of the complexity observed for shallow earthquakes resulted from reflections from the surface of the earth above the earthquake focus. For most shallow earthquakes, these reflections arrive within the first 35 seconds after P, and thus contribute to the complexity. When they are clearly resolvable, these reflections provide positive proof of depth, and thus make the "diagnostic aid" provided by complexity irrelevant. The "death knell" for complexity as an identification criterion, however, came a few months later in 1964 when the Soviets began testing underground at Novaya Zemlya, as will be described later in the words of one of its principal advocates.

The Limited Test Ban Treaty

In an attempt to break out of the unproductive atmosphere of the ENDC, President Kennedy and Prime Minister Macmillan offered to send senior representatives to Moscow to negotiate directly with Premier Khrushchev. In due course the offer was accepted, and on 15 July 1963 the talks began. The U.S. delegation was headed by W. Averill Harriman, a former ambassador to the Soviet Union and at that time the Undersecretary of State for Foreign Affairs. He was supported by an experienced staff of high-level officials of the Kennedy Administration, and by scientists, including Frank Press of Caltech's Seismological Laboratory. The British delegation was headed by Lord Hailsham, Minister of Science. Premier Khrushchev personally led the Soviet delegation. The U.S. had expected to negotiate a comprehensive test ban treaty, but it soon became evident that Khrushchev would not consider such a treaty and that the treaty would not cover underground tests so Frank Press returned home a few days later.

The treaty was concluded in ten days, initialed in Moscow on 25 July 1963, and formally signed on 5 August by the U.S Secretary of State and the Foreign Ministers of the U.K. and USSR. It banned nuclear tests in space, in the atmosphere, and underwater. National technical means were accepted by both sides as adequate for monitoring the treaty to an acceptable yield level. No restrictions were placed on underground testing except that radioactivity, if any should be vented, should not go beyond territorial boundaries of the testing country. The resulting treaty was very close in its provisions, and language to treaties that the U.S./U.K. had offered at least three times before.

If I seem to be giving only a terse account of this first formal testing limitation, it is because seismology played no active role in the negotiations even though seismic problems played a critical role in shaping the form of the treaty. Besides, there are excellent published accounts to which I can add nothing. I recommend Seaborg's (1981) book, among others.

Treaty Safeguards

President Kennedy submitted the treaty to the U.S. Senate to "advise and consent" on 8 August 1963. He summarized the arguments for the treaty in his transmittal letter, describing it as "a first step toward limiting the nuclear arms race." Hearings were duly held by the Senate, and expert testimony—both pro and con—was forcefully presented to the committees that reviewed the treaty.

Of particular importance to seismology was the testimony of General Maxwell Taylor, Chairman of the Joint Chiefs of Staff. Taylor offered a list of criteria used by the Chiefs in assessing the military impact of the treaty, and then he concluded that the effect on the military balance of power would be relatively minor if adequate safeguards were maintained. He then recommended four safeguards that the U.S. should establish. In brief, these included the maintenance of a vigorous program of underground nuclear tests, the maintenance of modern nuclear laboratories and programs, and the maintenance of facilities and resources to test promptly in the atmosphere should the USSR violate the terms of the treaty. The fourth, "Safeguard (d)," called for "The improvement of our capability, within feasible and practical limits, to monitor the terms of the treaty, to detect violations, and to maintain our knowledge of Sino-Soviet nuclear activities, capabilities, and achievements." The safeguards were accepted by the Senate, and endorsed by the President as U.S. national policy. The Senate voted in favor of the treaty by a wide margin on 24 September 1963. Safeguard (d) became the cornerstone of our future requests for funds for seismological research and development.

FOURTEEN

Some Reflections on Post-LTBT Developments

By mid-1963, three and one-half years of vigorous research had resulted in greatly improved knowledge concerning the generation and propagation of seismic waves from underground nuclear explosions, and improved methods for discriminating against earthquakes. It had not solved the basic problem of positive identification of explosions. Neither had four and one-half years of negotiations resulted in agreement by the Soviet Union to allow the inspection required to resolve doubt about some seismic events. The result was the Limited Test Ban Treaty of 1963.

Although a Comprehensive Nuclear Test Ban Treaty (CTBT) remained an explicit national goal during U.S. administrations following the Moscow Limited Test Ban Treaty (LTBT) of 1963, the goal has not been fully reached. Brief comments on the more active attempts related to that goal follow.

Treaty on the Non-Proliferation of Nuclear Weapons (NPT)

Early efforts to prevent the spread of nuclear weapons, beginning within a few months of the first use of the bomb, have been outlined in Chapter 4. These efforts were unsuccessful, and by the early 1960's there were four nuclear powers, to be joined by China in 1964.

By the mid-1960s it had been demonstrated that nuclear fuel could be used to generate electric power and nuclear power plants were in operation in several countries. It had also been found that the technology was not as difficult as first believed, and that uranium supplies were more abundant than previously thought. It was projected that, within a few decades, hundreds of nuclear power plants would be in operation, and in many countries. But along with electric power, plutonium would be created.

And plutonium can be separated from uranium by expensive but well-understood means, and potentially used for manufacturing weapons.

After lengthy negotiations, the NPT was completed and opened for signature in mid-1968. It was initially signed by the U.K., U.S., U.S.S.R., and 59 other countries. Briefly, it prohibits the transfer of nuclear weapons, and weapons technology, between nuclear powers and non-nuclear powers. It establishes safeguards to prevent diversion of peaceful nuclear activities into programs leading to weapons. It promotes peaceful uses of nuclear energy, and it obligates nuclear powers to share technology with non-nuclear countries. And it calls for further progress in arms control.

The Threshold Test Ban Treaty and the Peaceful Nuclear Explosions Treaty

There was little activity on test ban matters per se for the next decade after the signing of the LTBT, in part because of the priority given to the NPT and to limiting strategic missiles, but perhaps also because of U.S. preoccupation with the war in Viet Nam. Then, in the spring of 1974, under a presidency in deep trouble over the "Watergate affair," the U.S. abruptly reversed a long-standing policy against a threshold treaty. The new policy once again advocated a ban on underground explosions that produced signals larger than seismic magnitude 4.75, which would permit the U.S. to test explosions of the order of 100 kt in limited places at NTS, and would not require other seismic stations or on-site inspection inside the USSR.

In the interagency working group that developed details of this policy, I argued that the proposal would be unacceptable to the Soviets. After all, I had testified before Congress in an open session, and other U.S. scientists had published articles in unclassified journals on the relationship between yield and magnitude in the various rock types at NTS. Soviet seismologists could not fail to understand that the U.S. had already conducted an explosion somewhat larger than 100 kt that had produced teleseismic signals of magnitude 4.75. This was *Mississippi*, which had taken place on 5 October 1962 in deep, dry alluvium. No such deep, dry alluvium existed on Soviet test sites. Therefore a threshold expressed in magnitude units would likely limit them to tests a fraction of the yield of NTS explosions. It seemed unreasonable to me to expect the USSR to accept such a handicap.

State Department and ACDA officials, however, stated that Soviet Foreign Minister Andrei Gromyko had already agreed to a threshold expressed in terms of magnitude. And so the proposal was developed and meetings with the Soviets were scheduled. A team of thirteen experts and government agency representatives was hastily formed and dispatched to Moscow at the end of May to determine whether an agreed technical basis for a treaty could be reached. After flying overnight from Andrews Air force Base to Copenhagen, we were met by a Russian army navigator, who escorted us to Moscow. Once there, we were guided through immigration authorities and into military automobiles for a hair-raising high-speed drive down the middle of the highways and streets of Moscow to the Rossiya Hotel, where we were to stay as guests of the Russian government.

We assembled at the U.S. Embassy the following morning, and later attended meetings at the U.S.S.R.'s Atomic Energy Ministry. Our group was chaired by U.S. Ambassador Walter Stoessel; the Soviet group was chaired by Atomic Energy Deputy Minister I.D. Morokhov. After greetings and introductions, we began preliminary technical discussions. A few minutes into these discussions most of our U.S. group ran into a surprise: the Soviets signaled that the "units" to be specified for the threshold must be discussed, and in the next meeting proposed that the units should be yield—not magnitude. After a recess of a few days to allow the U.S. government to reformulate its policy, those of us who had not rushed back to Washington during the recess were ordered to negotiate a treaty with a threshold defined in terms of yield. At least implicitly, we were to complete the treaty in time for signature during a U.S./USSR summit meeting scheduled for early July, 1974.

Strong scientific foundations for verifying such a treaty did not exist. Seismologists at the meeting understood that seismic signal amplitudes from an explosion of specified yield depended on a number of factors: most prominently physical properties of the rock at the explosion site, size of the shot chamber, depth of burial, and propagational conditions along the paths between explosion and stations. A quantitative description of how these properties affected signal strength did not exist, however. We would simply have to make informed guesses, and bank on future research to improve yield estimation, or obtain empirical data from shots of known yield to calibrate each geophysically distinct rock formation at each test site.

The Threshold Test Ban Treaty (TTBT) was negotiated in the next four weeks. At an early stage in the talks it was established that verification of the threshold was a complex technical matter, and that it would require the exchange of data about geophysical characteristics of the sites to be used for testing. Working out the details of such data exchange was relegated to the scientists of each group, who met regularly in informal sessions. "Boiler-plate" language for the treaty, such as declaring both country's dedication to ending the nuclear arms race, and reaffirming their mutual commitment to a comprehensive test ban, were worked out by the two delegations in formal sessions.

As the scientists reached agreement on the information to be exchanged, it was reported to the formal sessions, where each item would be codified in the formal language of a protocol to the treaty. The treaty extended the LTBT by banning underground weapons tests larger than 150 kt, a compromise between U.S. (100 kt) and Soviet (200 kt) proposals. It called for verification by national means, but specified that certain geological and geophysical parameters of "testing areas" of the test sites would be described to the other side.

It also required that the yields, locations, and depths of two nuclear tests from each "geophysically distinct testing area" be provided to the other side for calibration purposes. Creating confidence in the accuracy of this—the most useful but also the most sensitive—information to be exchanged was left as a matter for politicians to solve, beyond the realm of science.

The treaty was signed by President Richard Nixon and General Secretary Leonid Brezhnev at a Summit meeting in Moscow on 3 July 1974 along with a protocol to the Anti Ballistic Missile Treaty. The signing ceremony was preceded by a private reception for the President and senior U.S. summit officials (and the remnants of the TTBT negotiating team) in a marble-walled "winter garden" area of the Kremlin. The signing ceremony itself was followed by a splendid celebration and reception in the St. George Hall of the Kremlin.[45] The celebration was attended by most of the diplomatic corps of Moscow, and it featured countless buckets of caviar, dozens of whole smoked sturgeons, bottomless bottles of excellent Georgian champagne, and music by both U.S. and Soviet military bands. Well, it was another foreign policy accomplishment for President Nixon, but not sufficient to rescue his presidency. He resigned five weeks later on 8 August 1974.

A companion treaty regulating underground nuclear explosions for peaceful purposes was negotiated in Moscow between October 1974 and April 1976. The Peaceful Nuclear Explosion Treaty (PNET) contained lengthy and elaborate provisions for verification. These included the right to send "designated personnel" with carefully defined equipment to the site to observe each peaceful explosion, and to measure the yield of explosions whose "aggregate yield"[46] exceeded 100 kt.

Neither of these treaties was ratified. Incoming president, Jimmy Carter, had criticized both treaties during the U.S. presidential campaign of 1976. He regarded both as diversions from the real goal, i.e., a CTBT, and he felt that the 150 kt threshold imposed no effective restraint on weapons development. He also argued that the peaceful nuclear explosion treaty "endorsed" such explosions, and offered a convenient "smoke-screen" for future proliferators to use to justify weapons tests (an excuse India had used when it first tested a "peaceful" nuclear device on 18 May 1974).

Comprehensive Test Ban Negotiations (Again)

Soon after taking office in 1977, President Carter proposed resuming trilateral negotiations (U.S., U.K, and USSR) on a CTBT. The other parties accepted, and negotiations began in Geneva in the fall of 1977. Meetings were held in consular facilities of the three countries. The principal issue, as always, was verification. Atmospheric and near-space tests were not a problem: by that time the U.S. had satellites in orbit scanning most of the earth with sensitive optical and electromagnetic-pulse sensors, and monitoring the upper atmosphere and near-space regions with gamma ray and neutron detectors as well. But enhancements to the seismic system were highly desirable, and the need for on-site inspection still existed to resolve unidentified seismic events. New thinking seemed needed on both aspects to avoid the sterile impasse of past years of negotiation.

On-site inspection was dealt with by adopting a concept introduced by Sweden years earlier. It went by the name "voluntary inspection." Under this concept, if a country, the U.S., for example, detected an event of concern to it on the territory of another country, say the USSR, the U.S. could request permission to inspect the region around the event. The USSR could either grant permission or reject the request. Under President Carter, the U.S. adopted the position that voluntary on-site inspection could be as effective in deterring cheating as mandatory inspection, if the consequence of rejecting a request were to be seen as a sufficiently serious matter, e.g., the requesting side might decide it must withdraw from the treaty.

Supplemental seismic stations were dealt with by adopting a concept of "National Seismic Stations" (NSS). As this concept was developed by the U.S., the system to be installed in the USSR would employ equipment built by the U.S. but "operated" by the USSR. Critical components like seismometers and amplifiers would be installed, under U.S. observation, in 500 foot deep boreholes, and they would contain tamper-detection devices, as well as devices to authenticate the digital data stream sent to the surface. There, the data would be broadcast via satellite transmission systems to the capitols of all parties to the treaty, where all seismic data could be read by everyone. However, only the U.S. and U.K. would have the "key" to the authentication processor at Soviet sites to determine whether the data had been tampered with. The Soviets would have the right to establish equivalent stations on territories of the U.S. and U.K.

As negotiations continued, and the prospects of actually agreeing on a treaty became more likely, the U.S. Department of Defense and the Atomic Energy Commission decided to work together to design and test a possible NSS system. In a memorandum of understanding between the two agencies, the AEC assumed the responsibilities for developing the secure seismic stations (based on DOD sensors) and associated communications, while the DOD would be responsible for the techniques and systems for receiving, analyzing, and storing the resulting data.

Sandia National Laboratories, as agent for the AEC, designed and built five such seismic stations, which were deployed at five locations in the U.S. and Canada, under the name "Regional Seismic Test Network" (RSTN). During the same period, ARPA, as the responsible agent for DOD, sponsored a design study by the Lincoln Laboratory of MIT of a system that could manage and process the almost unprecedented volume of digital seismic data. The RSTN data were broadcast via satellite to Sandia Laboratories in Albuquerque, New Mexico for engineering analysis, and to the Center for Seismic Studies in Arlington, Virginia, which would implement the Lincoln Laboratory design for a seismic analysis system.

In July 1979 a group of Soviet and British scientists came to the United States to be briefed on details of the NSS system in Washington, D.C. It was the last time I would meet with my aging colleague from the Conference of Experts and subsequent meetings, Mikhael Sadovsky. The meetings were cordial and conducted in a non-political manner, with social events in the evenings, including a reception in the Soviet Embassy. The visitors then proceeded to visit the Sandia Laboratory and a RSTN station established at the former "Geneva Station" site near McMinnville, Tennessee. The highlight of the trip for the Soviet scientists was reported to be a performance of the nearby Grand Ol' Opry.

Although there was substantial agreement on the verification measures, the treaty was never completed. My own involvement with test ban treaty and verification matters was decreasing at about this time under the pressure of other duties, so I have little depth of personal knowledge about the reasons for this failure. Three reasons reported by others include: disagreement between the UK and USSR on the number of NSS to be placed in former British Colonies; a U.S./USSR agreement to place their highest priorities on concluding the SALT II treaty; and growing concern about the wisdom of a permanent ban on all testing, at least within the U.S. At any rate, the Soviet invasion of Afghanistan in December 1979 dashed all hope of concluding either a SALT II treaty or a CTBT. Negotiations ended in November 1980, shortly after the election of President Ronald Reagan, and were not resumed during his administration.

Subsequent Negotiations

For completeness, I should report that in 1982 President Ronald Reagan decided not to resume negotiations on a CTBT until improvements were made in the verification measures of the existing TTBT and PNET. Protocols to each treaty, which contained the provisions on verification, were subsequently renegotiated and strengthened in the late 1980's. But public opposition to any form of testing was increasing in the U.S., and perhaps surprisingly to some, also in the USSR, where citizens were exercising newly found rights to criticize their government. In 1991 President Gorbachev announced a unilateral moratorium on Soviet testing and in 1992 the U.S. Congress passed a bill calling for a testing moratorium by the U.S., and requiring negotiations on a CTBT. In 1993 the new administration of President Bill Clinton began negotiations within the Conference on Disarmament, and a CTBT was successfully drafted and approved within the U.N. in 1996.

The treaty calls for the establishment of an International Monitoring System (IMS) with facilities for detecting seismic, infrasonic and hydroacoustic signals, and for collecting radionuclides in the atmosphere. Participating countries would "cooperate in an international exchange of data" from these facilities, and provide the data to an International Data Centre (IDC) for storage and processing. Processing by the IDC would be done "without prejudice to final judgments with regard to the nature of any event, which shall remain the responsibility of States Parties" to the treaty. The core capability of the seismic system will rest on a network of 50 stations equipped to transmit continuous, real-time data to the IDC, located in Vienna, Austria. This network is to be backed by 120 "auxiliary stations" from which data can be requested if needed. Sixty infrasonic and 80 radionuclide collection stations are also to be established. Eleven hydroacoustic stations are to be established, and there are provisions for on-site inspection. The treaty specifies that all 44 nuclear-capable countries must ratify before the treaty goes into effect—an unlikely happening in today's international climate. However, the IMS is well on its way to completion through the work of a Preparatory Commission.

President Clinton signed the treaty in September 1996. One year later the President sent the treaty to the U.S. Senate for its advice and consent; the Senate subsequently refused consent to ratification. The future of the treaty is uncertain, but adhering to international precedents, the signatories have so far respected the treaty's constraints, ratified or not.

What has changed?

At the time of writing, more than 160 countries have signed the CTBT, including the three nuclear powers that had previously been unable to reach agreement on verification issues since negotiations began in 1958.

We might ask what made agreement finally possible. The following sections address three possible areas of change.

Seismic Monitoring

Progress in seismic monitoring is an element, of course. Monitoring equipment and systems have become more reliable, although optimal sensitivity is still mostly dependent upon careful selection of quiet sites and the employment of instruments and installation methods designed to minimize noise—conditions that are not uniformly met by IMS stations. Seismologists have kept up with, and exploited, advances in digital technology and communications. The consequences are that seismic data from numerous worldwide stations are easily and rapidly available to monitoring centers, and the analyses of the data can employ sophisticated processes and be relatively thorough on each well-detected event. International cooperation in monitoring is a modern day fact of life, initiated through activities of the Conference on Disarmament and its technical arm, the Group of Scientific Experts (GSE). It would surely give a nation tempted to "cheat" on a test ban reason to think at least twice, knowing that much of the world may be watching for a violation. Vigorous research, especially in the U.S., which continued through more than four decades has significantly increased knowledge of the generation and propagation of seismic waves from earthquakes and explosions—the scientific foundations of identification.

Much remains to be done, however. Some of the identification criteria most relied upon are intrinsically empirical and statistical. Thus, in the absence of a clearly understood, sound physical basis, the effect of changes in source conditions cannot be predicted with confidence. Perhaps the most highly regarded of these is the $M_s{:}m_b$ criterion. There is good evidence that the criterion might fail for physical reasons at about $m_b = 4$ (several kt under coupling conditions similar to those at NTS). In any case, at well above that magnitude the surface waves are weak or undetectable at relevant distances for determining M_s. This leads to unreliable estimates of M_s, and hence unreliable discrimination.

A more recent discriminant, conceptually similar to $M_s{:}m_b$ in that it is based on the relative amplitudes of different types of waves, is applicable at smaller distances. It uses amplitudes of longitudinal and shear waves measured at frequencies higher than a few Hz. The developers of the discriminant have measured the ratio of P to S amplitudes, corrected for distance and propagation factors, from a population of earthquakes and another of explosions and tried to define a decision line that separates as many earthquakes as possible from the general population without including any explosions. This ratio has been found to be larger for explosions than for earthquakes at several test sites, as well as for the North Korean explosion of 9 October 2006 (Kim and Richards, 2007).

These methods can be applied to "screening out" a high proportion of the earthquakes, under the doubtful assumption that the explosion population is truly representative of all possible future explosions. The residual "unscreened" earthquakes are mostly lower magnitude events, and their numbers increase rapidly as magnitude decreases. However, exceptions have been reported well above m_b=4. The three PNE's in the Urals on 2 September 1969, 8 September 1969, and 25 June, 1970 (Marshall and Basham, 1972) and the Bukhara explosion of 30 September 1966 (Bolt, 1976) are examples. Reasons for these discrepancies might be understood in terms of the work of Stevens and Day (1985), who found that the differences observed between small earthquakes and explosions may be due to differences in such factors as depth and rock type in which these events occur, rather than to differences in source type. Experimental data demonstrating sensitivity of P waves to such factors have been reported by Murphy (1996) and Murphy et. al. (2001). Conceivably, such physical factors might be exploited purposefully to evade identification of a test. Statistics derived from past tests under normal test site conditions, no matter how overwhelming they may seem, do not address this possibility. Similar considerations apply to other criteria in this class.

Seismologists have had surprises before. We might pause to consider the fate of another empirical "criterion," highly touted in the United Kingdom in the early 1960's. In the words of one of the "complexity" criterion's principal advocates (Thirlaway, 1985):

From the accepted notions of [earthquake and explosion] source mechanisms, there were plausible reasons why earthquakes should generate more complex wavetrains than explosions. At first, these were not matched quantitatively when modeled and when the first large underground explosions in the Lower Paleozoic structure of Novaya Zemlya generated relatively complex P-wave trains at North American stations, the criterion was destroyed.

Beyond empirical/statistical criteria, there are criteria with relatively firm physical bases. The direction of first motion is one such criterion for identifying earthquakes, but is plagued with uncertainty in practice except for large events that produce signals recorded at very high signal-to-noise ratios. Similarly, the radiation pattern of surface waves is distinctively related to faulting motion. But as Brune and Pomeroy (1963) showed, an explosion may also excite such motion, leading to its misidentification as an earthquake if there is not intelligence or other independent information pointing to an explosive source. Even the most effective of the physically sound criteria for identifying earthquakes, determination of focal depth, should be viewed with caution, or even suspicion, when calculated from P-wave arrival times only. Unless the event is both very deep and recorded by numerous stations well-distributed in azimuth and distance, if it is not confirmed by telltale surface reflections (pP, sP), an event calculated to be deep may or may not be deep, even though the calculated confidence ellipse does not come close to the earth's surface.

Progress on such problems can be expected to be slow, having resisted more than forty years of intensive research. Additionally, by banning nuclear tests, we have forgone almost all opportunity to explore some of these matters experimentally. Nevertheless, approaches to solving such problems exist, and many excellent research scientists are still at work.

But the fundamentals of the seismological detection and identification problem have not changed. In a literal sense, a comprehensive nuclear test ban treaty is unverifiable. There will always be a threshold below which an underground explosion can not be detected by internationally deployed networks. This threshold may be to some extent adjustable at the discretion of a party desiring to test because of his ability to select his location and physical conditions near the shot point. Physical properties of the rock in which an explosion takes place have long been known to affect seismic signal strength by at least an order of magnitude, although the full range of rock types may not be available in some countries. Depth of burial is also a factor, with greater depths reducing signal strength, with the added benefit of reducing the possibility of accidental leakage of radioactivity. Decoupling, or partial decoupling by firing in a large chamber is also a possibility for reducing detectability of seismic waves. My opinion is that decoupled tests would probably be constrained by cost and practical considerations to explosions under ten kilotons, and be limited to technologically sophisticated nations. However the only qualified expert (L.F. Meade, see chapter 11) in cavity construction whose opinion I am aware of put the potential at much higher yields.

There will also be a higher threshold below which discrimination is ineffective because the positive discriminants like depth of focus and rarefactional first motion are based on subtleties of the seismic waveform that require higher signal-to-noise ratios than are required for mere detection. Seismological research has lowered both thresholds over the years, but they remain as an integral characteristic of seismic monitoring. Thus we see that a CTBT is, in reality, a threshold treaty in which the threshold for underground tests is rather loosely defined in terms of the capabilities of the seismic network employed, and by knowledge and capabilities of potential treaty violators.

This threshold is also influenced by the confidence level adopted by monitoring nations in interpreting the data—essentially a political judgment. Low confidence interpretations that can be argued among seismologists (and no doubt will be on any event of potential significance), may, nevertheless, provide some measure of deterrence. For example, some parties to the treaty might accept that detection by only a few stations would be adequate to deter cheating. In such cases, the location will typically be so uncertain as to be almost useless—certainly inadequate to find a concealed test—and identification by rigorous seismic criteria will be impossible. Yet some seismologists of the GSE and others still sometimes cite the 3-station detection threshold as a measure of the effectiveness of seismic networks—meaningful, perhaps, only for confirming an announced nuclear test.

But experience indicates that no responsible nation will accuse another of a violation of a nuclear test ban treaty without basing it on data and interpretations of the highest quality, preferably basing this interpretation on redundant data leading to a unique conclusion. It may be instructive to consider the mysterious event of 22 September 1979 in this connection. Two optical sensors on board a satellite detected a flash of light on that date that appeared to emanate from a location southeast of South Africa. Its intensity varied in a pattern characteristic of light from a nuclear explosion: a short intense flash followed by dimming and then a more sustained second flash. No known natural source could create such a signal from that region of the world. Hydroacoustic signals that many experts thought could have been generated by the event were detected, but they could not be linked to the event with certainty. No radioactivity was detected, but the search was relatively minimal because of the remoteness of the region from air bases accessible by properly equipped aircraft. There was some intelligence that both Israeli and South African military ships were operating out of South African ports, but the region was cloud-covered, so verifying this possibility with overhead photography was not possible.

Relations between the U.S. and South Africa were strained at the time, so scientists who were charged with evaluating the available data were made aware that a false charge against South Africa would be a serious matter. Slight difference had been noted in the signals from the two satellite sensors. It was possible to explain these differences by assuming that a paint chip off the aging satellite reflected sunlight to the two sensors, their separation on the satellite explaining the differences in the two signals. Thus, doubts were cast on the initial conclusion that a nuclear explosion had occurred, and U.S. policy makers were spared the painful task of accusing South Africa of testing.

As applied to concealed underground tests, we can conclude that mere opinions or loosely supported plausibility arguments—one way or the other—by eminent seismologists are unlikely to be persuasive in consequential cases. In the absence of a well-tested, definitive, explosion "signature," we should expect that the effective threshold for seismic monitoring against a clandestine test will be well above the size of the minimum detectable event, and that the population of events that cannot be shown conclusively to be earthquakes will be of considerable size.

Do small tests matter? I don't know, but I note that seven of the first 21 Soviet tests were under 3 kt. It has been argued that planned monitoring systems and capabilities preclude any significant weapons development program to proceed without being detected. Perhaps so, but I wonder if this thinking includes consideration of low yield tactical nuclear weapons secretly obtained by a technologically developed but devious country seeking an "ace-in-the-hole" to play against some real or imagined rival. North Korea comes to mind here, although its officials chose to boast (a bit prematurely) and bluster about it instead. And does any thinking person doubt Israel's nuclear capability, although no tests have been reported?

Complementary Techniques

Monitoring for underground nuclear tests does not rest exclusively on seismic systems, as the most recent tests (by India and Pakistan) have shown. Satellite imagery has sometimes been effective in detecting preparations for tests—although not necessarily recognized as such—and in locating the positions of tests with high accuracy after-the-fact when no attempt has been made to conceal the test. In such cases on-site inspection, if permitted, might be effective in establishing the nuclear origin of an event, if there were doubts about the event's true nature.

Without such precise location information, as might well be the case if a careful attempt were made to conceal a test, the Vela on-site inspection program clearly showed that the inspections are unlikely to be successful. Although research was conducted on an impressive number of physical, geological and radiological effects, and grew to more than $6 million annually by 1966, there were few (if any) promising leads left to investigate by 1969, and the program was closed. As summarized by Dr. Steven Lukasik, then Director of ARPA, in his final report on inspection to Congress in 1971:

1)   Research has shown that visual inspection and radiochemical
      analysis are the only useful techniques.

2) Deep burial of the explosion will prevent surface disturbances and seepage of radioactive gases to the surface.
3) Nevertheless, on-site inspection could be a deterrent because of the evader's fear of miscalculations and mistakes.
4) Search rates will probably be slow — Reconnaissance: 14 unit-days to find 90 to 100 sites in 250 km
   Detailed search: 8.4 unit-hours per site
5) Gas sampling is slow —
   3 Samples per site
   12 Samples per day per two-man team
6) Because search and sampling rates are low, accurate seismic location would be needed to reduce the size of the area to be inspected, and to insure that the epicenter lies in the area to be inspected.

The U.S. continued to argue for on-site inspection during the comprehensive nuclear test ban treaty negotiations of 1977-1980, although eventually agreeing to "voluntary" inspection. A wide range of technical methods were proposed for application during inspection, even though they had already been exposed by the research program to be ineffective against concealed tests. The 1996 CTBT also provides for on-site inspection, apparently as a "confidence building" measure as Marshall (1994) concluded, even though he agrees that inspection would be relatively ineffective against clandestine tests.

Political Developments

International political changes, unlike the slow changes in the effectiveness of scientific verification measures, have been profound since negotiations first began. The end of the cold war and the great opening up of the secretive former Communist countries have created — even for many former opponents of a CTBT — a climate of confidence that the nations that have the greatest technological capability to conduct clandestine nuclear tests would not do so, or indeed, may have little or no compelling motivation to do so.

More recent nuclear powers, and apparent aspirants to nuclear status, have not signed the CTBT, by and large. It may be that such countries do not view clandestine tests as in their interest — rather they may prefer a well-detected test as a demonstration of power. Thus, it seems to me that political changes have been the predominant factors that have brought the traditional nuclear powers to a cessation of testing. Whether the treaty and restraint by the major powers will encourage others to refrain from developing weapons — a principal rationale for the treaty — remains to be seen. Evidence of nuclear programs by North Korea, Libya, and Iran lead to questions about this premise.

However, the inverse side is clearly untenable: the major nuclear powers would have no moral ground for asking for restraint by others if they were to continue testing. Further, absent threats of the kind we faced in the 1940's or 50's, and given the numbers and undoubted diversity of the weapons we already have, it would be difficult to find ethical grounds for testing. Ratified or not, and subject to some verification uncertainties, it seems clear to me that adhering to the treaty is in our best interest.

FIFTEEN

A Few Final Words

I was fortunate to be drawn into the field of nuclear test detection seismology almost from its inception. I had not known that such work existed, much less that it would develop into such an important field—both for U.S. national security and for the science of seismology, itself. Instead, I had expected to work in some capacity in a university or in prospecting geophysics, as did most geophysicists before my time. Work in this emerging field proved to be both challenging and rewarding in many ways. A few highlights follow.

Perhaps foremost among the rewards has been recognition that the entire field of seismology was stimulated by research focused on nuclear test detection. Manifestations of this stimulation range from such simple measures as page counts in the Bulletin of the Seismological Society of America (about 450 in 1958 to more than 2,000 in 1968), to a listing of direct contributions to fundamental science. Among the latter I would list significantly increased accuracy of travel time curves (and hence improved knowledge of the structure of the earth's interior), and detection of short period P-wave reflections from the inner core (and hence strong evidence for a sharp boundary). Data from the Worldwide Standard Seismological Network put seismicity statistics on a truly consistent magnitude scale for the first time, and were also critical in establishing the existence of transform faults in zones of sea-floor spreading (a key piece of evidence in understanding plate tectonics). Perhaps the most important contribution was the major step in creating the technology base of modern seismology: the people, analytical tools, and knowledge, each greatly enhanced in the 1960s and 1970s.

Another important reward was the privilege of meeting and working with individuals of the highest caliber. I have given thumbnail sketches of some of the scientists that I knew rather well, and some I worked closely with for brief but intensive periods of time. I attended a number of meetings of the Committee of Principals, and especially during the Eisenhower administration, watched and listened as high ranking members of the government advocated sharply differing views on the wisdom or verifiability of test ban. I was sometimes asked for scientific advice or data—the latter, all too frequently woefully lacking in those early years. But I was able to see, first hand, how senior advisors hammered out disagreements to produce agreed policy recommendations, or, in a few cases, laid out issues for the President to decide. And I met many other able people drawn to aid their government in a time of great tension between East and West. I count several of my Soviet colleagues among those individuals, even though we differed on certain points.

There were countless scientific and technological challenges. By and large, those of a technological nature were more readily solved. Examples were the development of low-frequency, low-noise amplifiers; development of seismometers suitable for deployment in deep boreholes; development of long-distance data transmission systems; and conversion to digital analysis systems. Challenges of a more fundamental nature remain with us, in some cases. I would cite, as examples, discrimination between small earthquakes and explosions, and accurate determination of yields of underground explosions from remote seismic signals. Significant advances in understanding on both matters have been made over the years, and I would cite improved understanding of regional differences in the efficiency of propagation of seismic waves as one of these. It is clear in retrospect that many of the arguments of the Experts' Conference and TWG-II might have been avoided had either side known that Soviet experience with large explosions took place in regions of very efficient seismic wave propagation, while almost all of ours were in regions of much less efficiency.

A serious challenge of an entirely different kind, was coping with politically instructed and constrained Soviet scientists during TWG-II. Nothing in my prior years had prepared me for refusal by scientists to even consider careful scientific measurements based on their own merits. Our *conclusions* could be challenged, of course, but assertions that carefully made measurements were invalid because they did not conform to some "standard" that even the Soviets could not define was outside of my understanding of the methods of science. While it is true that each side negotiated within a political context, the Western scientists were never instructed on the science they should use during their negotiation tasks, nor to produce "scientific results" because the state required them. The experience was instructive to me in understanding the profound difference between attempts at serious rational analysis, and rationalization of a preconceived viewpoint (as happens all too often in politics). I suspect that some of the Soviet scientists were uncomfortable with their constraints, although a few seemed to relish their role.

Which brings me to a point I have been questioned on at times: guidance received from the Department of Defense. There were strong differences of opinion on the wisdom of, or need for, a nuclear test ban, especially during the Eisenhower administration. Secretary of State John Foster Dulles, and his successor, Christian Herter, strongly advocated a test ban, supported in this viewpoint by both of the President's Science Advisors, James Killian, Jr., and George Kistiakowsky. Secretaries of Defense Neil McElroy and Thomas Gates were opposed, supported by AEC Chairman Lewis Strauss and John McCone. Yet these latter two agencies were responsible for the research intended to remove barriers to a treaty. Cynics have suggested that these latter agencies may have attempted to slow the research, or even attempt to guide it toward problems, rather than solutions. This point of view is entirely mistaken. Senior officials of both agencies recognized that they were merely stewards of a *national* program that their President very much wanted to succeed. I am certain that the program was always fully supported at the top, both in the sharply divided Eisenhower administration, and in those that followed.

APPENDIX A

A Brief Introduction to Seismology

As originally defined, seismology is the science of earthquakes, and their causes and effects. Here, we concentrate on some of their effects: the vibrations of the ground that they generate. But explosions, as well as earthquakes, can cause such vibrations of the Earth. These vibrations travel away from he source as *elastic waves* of several types. Some waves travel through the interior of the Earth, and some around the Earth's surface. Those that travel through the Earth are known as *body waves*. Those that travel around the Earth are known as *surface waves*.

There are two types of body waves, longitudinal (or sound) waves and shear waves. Longitudinal waves travel with the greatest speed, and like the familiar audible sound waves in air, the vibrations of the Earth are back and forth in the direction of travel as the Earth (or air) is alternately compressed and rarefied. Shear waves travel a little more than half as fast, and like the motion of a long, straight garden hose when one end is shaken vigorously, the oscillations of the Earth are perpendicular to the direction in which the waves travel. Seismologists refer to these two types of waves as P (for "Primary," in the sense of first) and S (for "Secondary") waves.

Except very near the source, motions of the Earth from seismic waves are minuscule. P-wave motion from a 20 kiloton (equivalent to 20,000 tons of TNT) underground explosion can be smaller than one millionth of an inch at a distance of 2,000 miles. The time between successive crests of wave motion is called the period, and the waves we will be concerned with will usually be in the range of 0.3 seconds to several tens of seconds. These tiny and rather slow vibrations of the ground are detected by seismographs. Since the waves may be arriving from any direction (who knows where the next earthquake will take place?), and the waves may move the Earth in almost any direction, seismologists generally use instruments to record three components of motion, typically in the up-down, east-west, and north-south directions.

Earth scientists divide the Earth into four major concentric shells (Figure A.1). The uppermost is a thin rocky crust, ranging in thickness from several miles under oceans to thirty or more under high mountainous regions. The crust "floats" on a mantle, also made of rock, but more plastic than the crust. The mantle continues to a depth of 1800 miles—slightly denser and less than half of the Earth's radius. Below the mantle is a liquid core, slightly larger than Mars, believed to consist chiefly of molten nickel/iron. Within the liquid core is a solid inner core, probably also chiefly made of nickel and iron.

Because the Earth is round, the shortest path from an explosion site to a remote seismic detecting station is through the deep interior of the Earth. The speeds with which P and S waves travel increase with depth, and this causes these waves to follow curved paths that are concave upwards. As a result, they arrive at a distant station traveling steeply upward. For this reason, P-waves, which vibrate in the direction of the path, are best detected by seismographs that detect motion in a vertical direction (Figure A.1, ray arriving at 60°). S waves, on the other hand, vibrate in a direction perpendicular to the direction of travel, and are thus best detected on seismographs that detect motion in horizontal directions. P-waves and s-waves that are most detectable at distances of thousands of kilometers usually have periods in the range of 0.3 to 1.0 seconds. They are detected by short-period seismographs.

At each boundary within the Earth where the velocity changes, part of the energy of the wave is reflected, and part passes through. Light impinging on a glass window pane is a familiar example of this aspect of wave motion—some light is reflected while most passes through. For seismic waves it is a bit more complicated, however: for when a P-wave impinges on a boundary, both a P-wave and an S-wave will usually be reflected, as well as pass through. Similarly for S-waves. Since the Earth is multilayered, this means that a very simple source—say a single explosive pulse—will result in multiple reflected and transmitted waves traveling within the Earth.

In addition to the body waves, there are also the surface waves that travel around the Earth. The most important surface waves are known as Rayleigh Waves and Love Waves, each named for the mathematical physicist who developed the theory explaining it. Rayleigh waves are similar to the familiar waves that propagate on the surface of water in that the water moves in both the vertical direction and in the direction of travel. Love waves, on the other hand, produce no vertical motion but cause the Earth to shake from side to side perpendicular to their direction of travel. At distances of thousands of kilometers, surface waves are usually most detectable at periods of about 10 to 30 seconds. They are detected by long-period seismographs.

APPENDIX B

Pasechnik and the Special Monitoring Service

This appendix contains selected historical events inside the Soviet Union related to the development of seismic methods for detecting nuclear explosions. It is based on personal correspondence with Dzhamil Sultanov and Alexei P. Vasiliev, as well as a copy of a talk by the latter. Dzhamil Sultanov was a colleague of I.P. Pasechnik in establishing the scientific basis for detection of explosions in the early years of the Soviet research program. A.P. Vasiliev is a retired member of the Special Monitoring Service, and the primary historian and biographer of the SMS. I am indebted to both.

According to Vasiliev (1999), the Soviet Union's interest in the long range detection of nuclear explosions was triggered by their investigation of the information "leak" on the occurrence of the first Soviet nuclear explosion, which we called *Joe-1*. Vasiliev writes that "this information leak was absolutely surprising for the Soviet Atomic Project leaders, as the Project was a matter of the highest level secrecy in the country." Furthermore, the Soviets had detected radioactive fall-out only out to 750 km from the Semipalatinsk test site, and they believed no detectable radioactivity had passed beyond their national borders.

Serious scientific studies of long range detection began in 1951 within the Laboratory of Measurement Devices of the Academy of Sciences of the USSR. This work, on transmission of nuclear debris through the atmosphere, was under the direction of the renowned physicist Igor V. Kurchatov, Scientific Director of the Soviet Atomic Project. That same year G. A. Gamburtsev, Director of the Geophysical Institute of the USSR Academy of Sciences, began to study seismic waves created by nuclear explosions.

The first seismic recording of a nuclear explosion by the Soviets was made on 24 September 1951. The event was *Joe-3*, a 38 kt shot on a 30 m tower at the Semipalatinsk test site. It was recorded at a distance of 700 km near the village of Borovoye, Kazakhstan, using a high-frequency seismograph designed by Gamburtsev. Among his assistants was a scientist whom we would later meet during the Conference of Experts, Ivan P. Pasechnik. The new instrument successfully recorded the shot, but as reported by Vasiliev, in the "aura of top secrecy around the Atomic Project" the seismogram was ordered to be burned by the KGB (Committee of State Security of the Soviet Union).

In February 1954 a classified "Seismometrical" laboratory was established within the Geophysical Institute of the Academy of Sciences, with Pasechnik as its head and Dzhamil Sultanov as one of its first scientific staff members. Its mission was to develop seismic methods for monitoring nuclear tests. One of the laboratory's early accomplishments was the establishment of sensitive seismic stations at Mikhnevo (near Moscow) and Kuldar (in the Far East). In 1956-1957 similar stations were deployed in Antarctica. These stations were initially equipped with SVK-M and SGK-M seismographs, designed by Pasechnik and his associates to have peak magnification at 1 Hz. Pasechnik and his colleagues also developed other special seismic sensors, the USF (universal horizontal and vertical) and the SDF (for use in a borehole), as well as microbarographs.

By 1954, both airborne and ground-based methods for collecting radioactive debris had been studied and long range detection of acoustic and electromagnetic pulse (EMP) signals had been demonstrated on *Joe-4*, the first thermonuclear test at the Semipalatinsk Test Site. A Service for Special Observations (SSO) was created in 1954 within the GRU (the Chief Intelligence Directorate of the General Staff of the Soviet Union). The new service was responsible for testing the possibility of detecting EMP using existing military communications systems such as those used for intercepting radio traffic for intelligence collection, as well as to develop new techniques for emp detection. The SSO was commanded by GRU Col. Alexander I. Ustyumenko, whom we later met as Dr. Ustyumenko in Geneva.

Experimental work continued on the various methods, with active measurement programs only during periods of nuclear testing operations at Soviet and U.S test sites. The SSO did not yet have a mission of around-the-clock monitoring for potential clandestine nuclear tests, as we would later surmise at the Conference of Experts. Seven tests of the 1956 series at Bikini were detected seismically in the USSR, and yield estimates were made by measuring the amplitudes of surface waves, according to Vasiliev.

By 1957 the theoretical and experimental studies had been completed by Soviet scientists, primarily within several institutes of the Academy of Sciences, that established the foundations of a long range detection system. It was to be based on seismic, acoustic and EMP detection, and on the collection of airborne radioactive debris. During that same year, the Soviets began a "step-by-step transfer from experimental studies to the [status of] real time nuclear explosion monitoring."

Results of this research and development were published by the Institute of Atomic Energy of the USSR's Academy of Sciences in a two-volume work titled "Development of the Nuclear Explosion Long Range Detection System." This work was awarded the Lenin's Prize in a competition among classified scientific works for 1958. Two of the authors who received this high award were I.P. Pasechnik and A. I. Ustyumenko.

Responsibility for implementing the developing monitoring system was assigned by the Soviet government to a Service for Nuclear Monitoring within the Ministry for Defense on 13 May 1958, six weeks before the Conference of Experts. Shortly afterward, Colonel Ustyumenko was promoted to the position of Commander of the Service. In 1960 it became the modern day Special Monitoring Service (SMS), the Soviet counterpart of the United States' Air Force Technical Applications Center (AFTAC).

While the SMS had full responsibility for operational aspects of monitoring foreign nuclear tests, responsibility for research and academic activities associated with seismic detection remained with Pasechnik's laboratory within the Academy of Sciences. A close relationship between the Service and the laboratory was maintained for many years. The first openly acknowledged workers in the field of nuclear monitoring, whom I met in Moscow in 1974 while participating in Threshold Test Ban Treaty negotiations, described Pasechnik as their mentor. Initial work on seismometer arrays and digital recording began in 1961 at a research facility established near Boroyoye. Pasechnik's laboratory was also responsible for research on generation and propagation of seismic waves, and the development of methods of discrimination between earthquakes and explosions. It included a large program to measure seismic waves from peaceful nuclear and chemical explosions at test sites and elsewhere throughout the Soviet Union from 1958 to 1990. This program was under the direction of Dzhamil Sultanov, now the dean of Russian nuclear detection seismologists.

APPENDIX C

Epicentral Location by Stereographic Maps

We used a polar stereographic projection (projection on an "equatorial" plane of each point of latitude and longitude along a line to the opposite pole). The key feature of the polar stereographic projection is that any circle on the earth projects as a circle on the map. We were, therefore, able to construct a chart for each station with circles drawn on it at distances corresponding to each 20 seconds of travel time of P-waves out to about 14 minutes and 20 seconds, corresponding to an epicentral distance of about 103°. Charts were constructed on poster board, about 24 x 30 inches, centered on the U.S.S.R. and China

Using the trial epicenter and origin time based on the estimates from the Wyoming tripartite station network, the analyst could identify the proper travel-time circles on the map. Portions of the circle were traced onto a transparent overlay along with basic map coordinates. This was repeated for each station. If all circles intersected in a point on the overlay, the analyst had located the epicenter. If not, the trial location and origin time were adjusted somewhat and the process repeated until convergence was achieved. It was an art, admittedly, but one in which the analysts became highly skilled.

The basic travel-time charts were also used to construct auxiliary charts showing arrival time differences between each pair of stations. The method is similar to the Loran radio navigation system used during World War II. The observed difference in times of P-waves at stations A and B meant that the event occurred somewhere along a hyperbola-like curve on the chart; a similar curve for stations B and C will intersect the first at two points, in general (although one may often be ruled out for various reasons. The curve for a third station pair, C and D, will go through only one of the two intersections, thus providing an even more accurate trial epicenter than available from the Wyoming Tripartite Network.

APPENDIX D

Epicentral Location Error

In principle, the four focal parameters defining the occurrences of a seismic event (latitude, longitude, depth and origin time) may be calculated from the arrival times of P-waves at four stations. (Three stations may suffice if clear S-waves are also detected, but this is unusual for small events unless the event occurs within a few hundred km of a station—improbable for the networks being considered.) But we might ask how accurate will the locations be?

The Experts' Conference concluded that "making use of the data of several surrounding seismic stations, the area within which an epicenter is localized can be assessed as approximately 100-200 square kilometers." This corresponds to the area of a circle 5.6 - 7.9 km in radius, roughly the distance a P-wave travels in one second at epicentral distances up to about 1,000 km. And that fact, incidentally, summarizes most of the logic behind the estimate.

The availability of high speed computers gave seismologists the means to make more sophisticated estimates, and data from *Gnome* and numerous other explosions allowed these methods to be tested in the early to mid 1960s.

Early work by E.A. Flinn showed that it was possible to compute the four-dimensional ellipsoid which describes the joint confidence region for the four focal parameters. Using this method the orientations and lengths of the axes of an ellipse can be calculated, wherein it can be estimated with a specified probability that the epicenter lies.

Applying this method to observed P-wave arrival times from large explosions and earthquakes Romney and Flinn (1963) showed that, at the 75% confidence level, achieving epicentral uncertainties as small as 10 km (i.e. 300 km2) requires data from about 30 stations. Uncertainties increase rapidly as the number of stations decreases, exceeding 40 km for fewer than six stations (see Figure D.1). Similar results have been found experimentally in a more recent study by Sweeney (1996).

However, the actual error may be larger; the estimate assumes that the travel time table used in the calculations are an accurate representation of the actual times taken for P-waves to travel from the source to each station. This is unlikely to be the case in the real, heterogeneous earth. As noted in Chapter 11, large deviations from travel time tables exist, and can bias the calculated epicentral location. In an exceptionally well calibrated region Thurber et. al., (1997) showed that most calculated locations were outside of the 95 percent confidence ellipses.

In recognition of the uncertainties of seismic locations, the comprehensive Test Ban Treaty of 1996 calls for on-site inspection rights in up to 1,000 square kilometers.

APPENDIX E

Hardtack-II Amplitude and Magnitude Data

Measured P-wave amplitudes for *Rainier*, *Logan*, and *Blanca* are given in the tables referenced here. The maximum vertical amplitudes in the first three cycles of the P-waves, divided by the period, t, are given in Table E.1. this quantity, a/t, is used in computing magnitudes and is more directly related to energy than is the displacement. For convenience, and since it is not a true velocity measure, a/t will simply be called "amplitude" in this appendix.

The local magnitude, M_L, for *Rainier* and the Hardtack shots are given in Table E.2.

Teleseismic magnitudes, m_b for *Logan* and *Blanca* are given in Table E.3. Estimates indicated as <4.5 are based on noise levels.

APPENDIX F

The Large Aperture Seismic Array (LASA)

LASA was intended to be the culmination of the Vela Program's effort in array design. It was perhaps best described as a 200 km diameter array of 21 more conventionally sized 7 km subarrays. Its core sensors were 25 short-period vertical seismometers in each subarray, each seismometer installed in a 200-foot-deep borehole. Thus, functioning as a 525 element array, it incorporated features of both small noise-reducing arrays and larger velocity filtering arrays. Each subarray was additionally equipped with two short period horizontal seismometers and a three-component set of long period seismometers.

First proposed in 1964, it was rapidly constructed near Billings, Montana, and went into operation in mid-1965. Construction in such a short time, and mainly over a stormy winter in windswept central Montana, was a model of both good engineering and good management. The station operated as a research tool with all 630 sensors until 1970, and in a reduced configuration until 1978.

The major seismological premise underlying the subarray design was that coherence of seismic noise could be exploited to improve signal-to-noise ratio (SNR) in the P-wave signal band by more than the square root of the number of detectors, n. Initial reports of high gains in SNR proved to be based on noise reduction at longer periods *outside* of the signal band. Such out-of-band noise is normally removed by conventional frequency filtering. When SNR measurements were confined to the P-wave signal band, the array gain in SNR was less than the square root of n. Two factors contributed to this disappointing result. The first was that the noise within the subarrays was far more intractable than hypothesized, and the noise was reduced by less than the square root of 25. The second was that the P-wave signals were not fully coherent over the large distances between subarrays, and the combined signal amplitudes were less than n times the individual amplitudes. In the end, the LASA detection threshold for P-waves proved to be about magnitude 3.8, which had already been achieved by small arrays constructed in carefully selected quiet sites.

On the other hand, the LASA experiment advanced array design and engineering, and the array's data supported numerous other seismological studies. Although not the first use of digital data in seismology, pioneering work was done in the large-scale acquisition, use, and management of digital data. Velocity filtering capabilities, as applied to signals, were shown to be excellent, and were exploited to give improved understanding of the nature of the P-wave coda, to cite an example of basic seismological studies the array supported. I believe the data are still in use to support research.

APPENDIX G

Horizontally Polarized Short Period Shear Waves

The observation of large shear waves from underground nuclear explosions, and especially the horizontally polarized shear waves, came as a surprise to most seismologists. Beginning with *Rainier*, it had been found that short period (one to several Hz) S-waves were essentially as large, relative to the P-waves, as those from earthquakes. (This may not hold true at higher frequencies, as outlined in Chapter 14.) Initially, it was hypothesized that these waves were formed through conversion of other wave types as they traveled through the heterogeneous crust between the source and the stations. Another early conjecture was that they occurred as a result of slippage along joints or faults in the rock in which the explosion took place. The fact that the earliest U.S. underground explosions took place near the edge of a mesa was also advanced as an explanation for short period S waves, and when Love waves were subsequently noted from Logan and Blanca, for these waves as well.

Conversion between wave types was soon shown to be insufficient to explain the observations. If Rayleigh waves, for example, could be converted to Love waves by inhomogeneities along the path of propagation, why were Rayleigh waves, but not Love waves, found from atmospheric explosions? Attention thus focused on near-source effects, and it was soon found that joints, faults and layering of rocks had an important effect on the shape of the cavity created by the explosion (Short, 1961) and , by inference, probably on the amount of energy in shear waves. Furthermore, Wright and Carpenter (1962) showed theoretically that P-waves attenuate much more rapidly in the inelastic region near the explosion than do shear waves. Accordingly, even small amounts of shear near the source can become large relative to P as the motion is transmitted to the distance where true elastic waves exist. Not conclusive, perhaps, but strong reason to explain why discrimination based on amplitudes of shear waves might not be completely diagnostic.

Notes

Notes to Chapter One

[1] A common practice in seismology is to report distances in kilometers out to 1,000-2,000 km, and in degrees measured along great circles between source and station, at greater distances. 90_ is 10,000 km.

[2] $\text{Log } E = 11.3 + 1.8 \text{ M}$

[3] Dr. John Adkins, Office of Naval Research; Dr. Perry Byerly, University of California; Dr. Maurice Ewing, Columbia University; Dr. Norman Haskell, Air Force Cambridge Research Center; Dr. Frank Neumann, Dr. Dean Carder, USC&GS; Father James B. MacElwane, S.J., St. Louis University; Dr. Beno Gutenberg, Dr. Hugo Benioff, Dr. Charles Richter, Caltech; Dr. Louis Slichter, University of California, Los Angeles.

Notes to Chapter Two

[4] During World War II, the U.S. Navy supported research to determine whether microseisms could be used to track hurricanes. Methods were developed to determine the direction of travel of these storm-related waves in the hope that a hurricane could be located by using bearings from several seismic stations. (Ramirez, 1953). The method did not give accurate locations, and the research was abandoned.

[5] Spiral-4 cable was a shielded, four-conductor cable used by the Army Signal Corps for telephones in field installations. It was available in large quantities as surplus from World War II.

[6] Billy Brooks, John Coleman, Michael Flynn, Braden Leichliter, Howard Peterson, Carl Romney, and James Zivney.

[7] U-2 overflights of the USSR began 4 July 1956. I have no idea when this particular flight took place.

Notes to Chapter Three

[8] Pietro Caloi had earlier reported observations of PKiKP, but his published seismograms were unconvincing.

[9] $\text{Log } E = 9.4 + 2.14 \text{ M}_L - 0.054 \text{ ML}^2$ This notation for magnitude, M_L, became used to indicate a calculation based on the amplitudes measured by Wood-Anderson Torsion Seismometers at relatively local stations. It is the original magnitude defined by Professor Charles F. Richter of Caltech.

Notes to Chapter Four

[10] The International Geophysical Year was a period of coordinated worldwide measurements of key properties of the atmosphere, oceans, and solid earth. It was timed to coincide with a sunspot maximum in 1957-58, and thus with increased auroral, magnetic, and other sun-induced activities. Complete sets of the data were assembled at three World Data Centers, and were freely available to scientists of both East and West.

[11] Speculation at the time was that the object was either tumbling or yawing in some regular way to cause the reflected sunlight to fluctuate.

[12] Included in Reference List, 1954. "Class A" are the largest, best-recorded earthquakes. "Shallow" meant a depth of 60 km or less.

Notes to Chapter Five

[13] There were a number of other advisors, some of whom were present only briefly. Those who played sustained roles included Spurgeon Keeny, Office of the Special Assistant to the President for Science and Technology; J. Carson Mark, Director of the Theoretical Division, Los Alamos Laboratory; George B. Olmstead, Assistant Technical Director, AFOAT-1; Herbert "Pete" Scoville, Jr., Central Intelligence Agency; and Anthony L. Turkevich, Enrico Fermi Institute for Nuclear Studies, University of Chicago.

[14] An Aide-Memoire from the Soviet Foreign Ministry to the U.S. Embassy in Geneva on June 13, 1958 had conditionally accepted an experts' conference, and named its eight formally selected experts. Several of those were known to us, others were almost unknown, and none of numerous advisors were identified in advance.

[15] The use of the word "controlling," where idiomatic English would have used a more passive word like "monitoring," reflects only one of numerous cases where we had to struggle with different meanings of the same words in English and Russian.

[16] Dr. Ernest O. Lawrence also returned to the United States in mid July. He had attended the Conference knowing that he was seriously ill, and he returned to obtain medical treatment. He died as a result of surgery not long afterward.

[17] This is also implicit in the earthquake magnitude scales of Gutenberg and Richter, and consistent with our measurements of P-waves from NTS and Pacific atmospheric shots.

[18] $A=CQ1/3/D$, where Q is yield and D is distance.

[19] "SK" was the notation for the standard Kirnos seismograph; "SVK" and "SGK" identify the instrument as a vertical or horizontal component, respectively.

Notes to Chapter Seven

[20] Later estimated at 22 kt, but it makes little difference for this discussion

[21] Standard deviation of a single observation.

Notes to Chapter Eight

[22] In seismological circles, this document on research is widely known as the "Berkner Panel Report," rather than the lesser known classified reports.

[23] The decision to assign the program to ARPA was influenced by the government's decision to implement not only the Berkner Panel's proposed seismic research, but also research on detection of high altitude and near-space explosions and involving satellite-bourne sensors. The latter research had been recommended by another panel chaired by Dr. W. Panofsky, working at about the same time as the Berkner Panel.

[24] As previously noted, ARPA was formed to manage Department of Defense space activities. It became their practice to name major programs after constellations. Vela is a sub-constellation of Argo; it represents the sail of the Argonaut's vessel.

Notes to Chapter Nine

[25] These numbers were believed to be uncertain by at least a factor of two. Thus 300 is a shorthand for 150-600.

[26] Other members were: Dr. Hans Bethe; Dr. Harold Brown; Dr. Richard Foose, Stanford Research Institute; Dr. Richard Garwin, IBM; Mr. Spurgeon Keeney, Office of the Special Assistant to the President for Science and Technology; Dr. Albert Latter, Rand Corporation; Dr. J. Carson Mark, Los Alamos Scientific Laboratory; Dr. John Tukey, Princeton University and Bell Telephone Laboratory; Dr Anthony Turkerich, University of Chicago.

[27] Outside of the politically charged atmosphere of TWG-II, eminent geophysicists (Griggs and Press, 1961) would later write of our experiment: "Romney's account of travel-time and amplitude determination across the United States from the Hardtack II shots will stand as a classic in seismology. Never before have such extensive and accurate records been obtained on a continental scale."

[28] Tukey was a witty and affable man who did pioneering work in the field of digital analysis. He was credited with coining the term "software" to describe processes that run on computer hardware, and "bit" for binary digit, the basic unit of digital information. Later, with J. W. Cooley, he developed the "Fast Fourier" transform algorithm, a foundation of modern waveform analysis.

[29] Their notation for mb differed from that used here, and differed among their several 1956 publications as well. I've tried to be consistent with current notations.

[30] $m_b = 1.7 + 0.8 \, M_L - 0.01 \, M_L 2$

[31] The Cowboy experiments were conducted in a salt mine near Winnfield, Louisiana. Decoupled and well-tamped chemical explosions ranging from 20 to almost 2000 pounds were carried out between 14 December 1959 and 4 March 1960. Published reports in 1960, too late for TWG-II, reported a decoupling factor of approximately 100 for shots in cavities as compared to tamped shots, both in salt.

[32] Ieuan Maddock played a prominent role in nuclear test detection for many years. He had been present for the first British nuclear test off the northwest cost of

Australia at Monte Bello Island. He had announced the "countdown" leading up to detonation. For this, he was known to friends as "the Count of Monte Bello."

Notes to Chapter Ten

[33] Other members of the panel were: Austin W. Betts, ARPA, Francis G. Blake, Jr., California Research Corp., Maurice Ewing, Columbia University, Philip J. Farley, Dept. of State, Gerald W. Johnson, Lawrence Radiation Laboratory, Richard Latter, Rand Corp., Julius P. Molnar, Sandia Corp., William E. Ogle, Los Alamos Scientific Laboratories, Frank Press, Cal Tech, Carl F. Romney, AFTAC, Glenn A. Smith, CIA, Alfred D. Starbird, AEC, John W. Tukey, Princeton University.

[34] Other members of the U.S. group were: Mr. Carlton M. Beyer, Dept. of Defense (ARPA), Dr. Gerald Johnson, Lawrence Radiation Laboratory, Mr. Spurgeon M. Keeny, President's Science Advisory Committee, Dr. Richard Latter, Rand Corporation, Dr. Donald H. Rock, Dept. of Defense (AFTAC), Mr. Robert C. Scheid, Dept. of Defense, Dr. M. Carl Walske, Atomic Energy Commission.

[35] With logic understandable only to bureaucrats, the programs were named VELA UNIFORM (for underground), VELA HOTEL (for space-based sensors) and VELA SIERRA (for surface based detection of high altitude tests).

[36] I don't believe the Panel members expected to solve the basic problems in such a short time. Much of the focus was on demonstrating "quick-fixes" to improve the Experts' systems. However, we hoped that early successes would establish the value of the work, and result in a sustained program.

Notes to Chapter Eleven

[37] Other members of the ad hoc group were: Dr. Hugo Benioff, California Institute of Technology; Dr. John Crawford, Continental Oil Company; Dr. Freeman Dyson, Institute for Defense Analysis; Dr. Val Fitch, Princeton University; Dr. Roland Herbst, Lawrence Radiation Laboratory; Dr. Wolfgang Panofsky, Stanford University; Dr. Alan Peterson, Stanford Research Institute; Dr. Frank Press, California Institute of Technology; Dr. John Tukey, Bell Telephone Laboratories; Dr. Kenneth Watson, Lawrence Radiation Laboratory; and Dr. Jack Oliver, Columbia University.

[38] The thinking had been based on elastic theory, which predicted that seismic amplitude at great distances would vary inversely as the rigidity times the speed of the P-waves. It was also believed that the elastic radius would be determined by lithostatic pressure due to the weight of the rock above, rather than to the length or other mechanical properties of the rock; as a corollary this radius would vary little from one rock type to another since densities of most rocks used for testing are close to one another. Accordingly, a shot in faster, more rigid, salt should produce smaller signals than a shot in tuff or alluvium.

[39] Stretch computers were the supercomputers of the time.

Notes to Chapter Twelve

[40] Brazil, Burma, Ethiopia, India, Mexico, Nigeria, Sweden, United Arab Republic

[41] Its seismic magnitude was about 5.2, based on measurements out to about 90°. As late as August 1968 this test was reported authoritatively to be the first French underground nuclear test explosion (SIPRI, 1968) In fact, the first French underground explosion had taken place the previous year on 7 November, 1961, about four weeks after the first Soviet underground test. Like the first Soviet test, it was a low-yield test (yield still undisclosed), and both were little noticed at the time outside of AFTAC. The SIPRI report, prepared by eminent seismologists from ten countries, was intended as a review of the state-of-the-art in monitoring underground nuclear explosions. Appendix I, "Seismologically important nuclear explosions," identifies the large second underground tests of both Russia and France as first detonations, apparently without knowledge of earlier low yield tests by each country. One of the seismologists, I.P. Pasechnik, was certainly aware of the earlier Soviet test, but was probably under security restrictions not to divulge a test apparently missed by the West. Marcus Båth (1962) had reported that he had been informed of the first Soviet underground test by the USC&GS and he reported data from the single station (Skalstugen) in Sweden that had recorded a signal. His description of the large 1 May 1962 French test implies that he was aware that there had been an earlier test (informed by the USC&GS?) but he appeared not to know the date.

[42] The "Shot Report" produced by the LRSM program, a compendium of measurements of times and amplitudes of signals recorded from the event, gave an "average" magnitude of 4.9, but the report included measurements at distances less than 16°, which are known to inflate the average.

[43] Elastic theory predicts that seismic motions from P-waves of a given energy will be larger in lower velocity media than in a high velocity medium such as granite.

Notes to Chapter Thirteen

[44] In its tenth annual meeting in London in September 1962, the "Pugwash" Conference on Science and World Affairs , a "Statement on Test Detection" was issued. The conference took its name from the location (Pugwash, Nova Scotia) of the summer home of one of its sponsors, U.S. Industrialist Cyrus Eaton, where the first conference was held. The agreed statement had been signed by three American and three Russian scientists. They presented a naive concept of "black boxes" handed over to Soviet scientists,, to be placed by them at agreed locations. They indicated that such unmanned, automatic seismic stations would produce sufficient reliable data that the control commission would need to "request" very few on-site inspections. The three Americans were all physicists who knew little or nothing about seismology, as was apparent from the proposal.

Notes to Chapter Fourteen

[45] St. George is the patron saint of Russia.

[46] The sum of the yields of individual devices when several were to be detonated simultaneously, e.g., several shots in a line to excavate a canal.

Figures

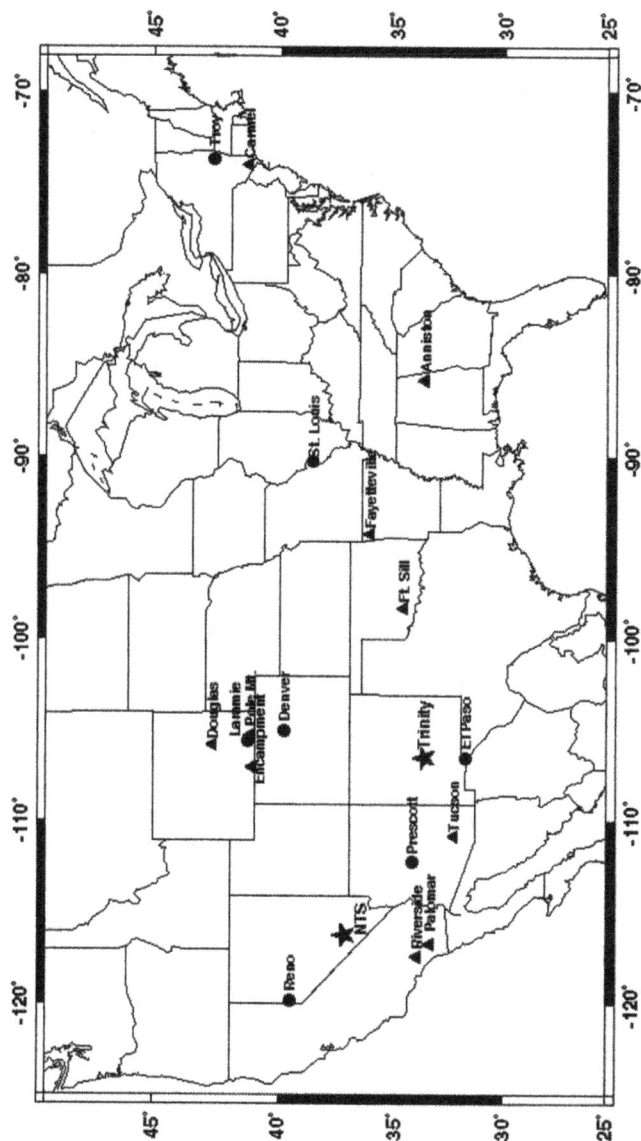

Figure 1.1 Location of Trinity and Nevada Test Site (NTS). Cities and seismic station localities mentioned in first three chapters are also shown.

Figure 1.2 P-waves from Bikini Baker as recorded by a Benioff seismograph. Time increases from left to right; the square deflection to the left marks the beginning of a minute. Actual vertical motion of the earth is about 1/1,000 millimeter. Wave period is about one second.

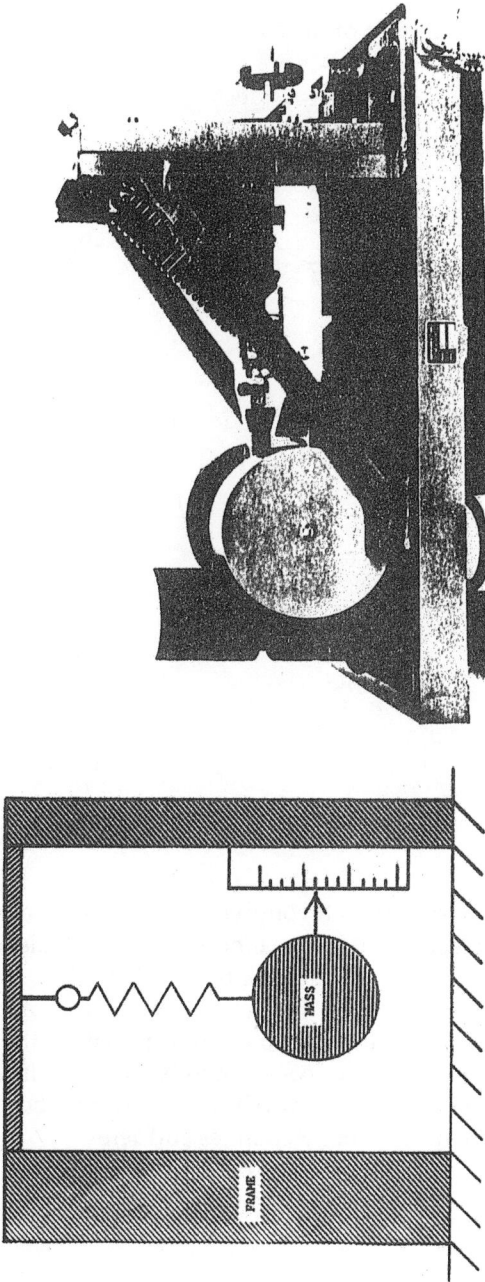

Figure 2.1 Left: Schematic of a vertical seismometer, showing spring-suspended mass. Motion of the earth relative to the inertially stable mass is recorded.
Right: Actual long period instrument has a mass mounted on the left end of a horizontal hinged beam that is supported by a heavy spring. An annular coil, mounted on the mass, moves through the field of a magnet mounted on the frame to generate an electric output.

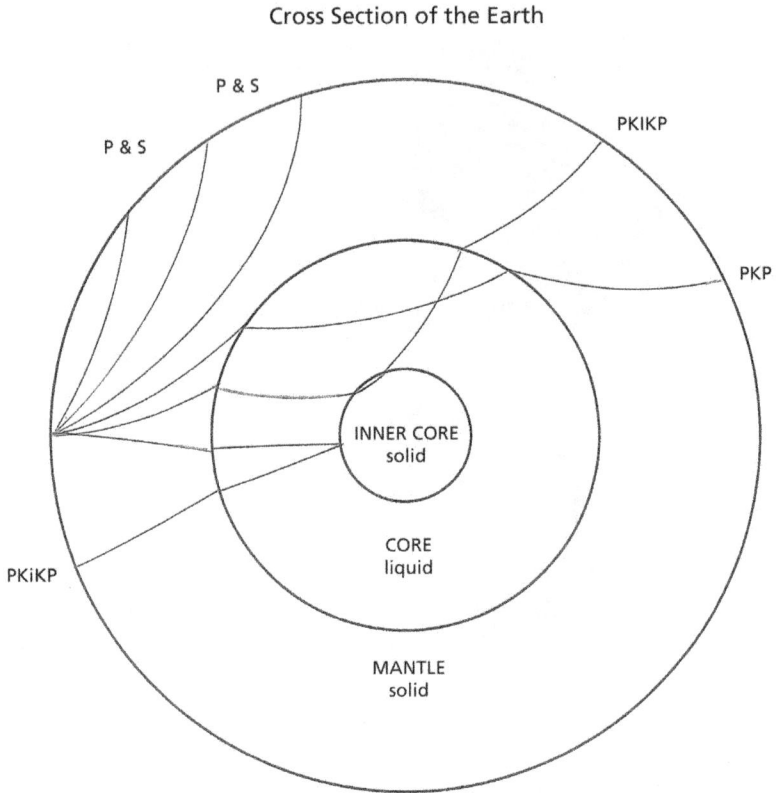

Figure 3.1 Schematic seismic ray paths through the earth. As the angle of departure from the source steepens, the rays emerge at greater distances. An increase in velocity with depth in the mantle bends both P- and S-waves upward toward the surface. Rays impinging on the core are bent sharply downward because of a sudden drop in velocity within the core. Still steeper rays impinge upon the inner core, where an increase in velocity sends them upward again. Waves may also be reflected from boundaries of the core or inner core where velocity changes suddenly.

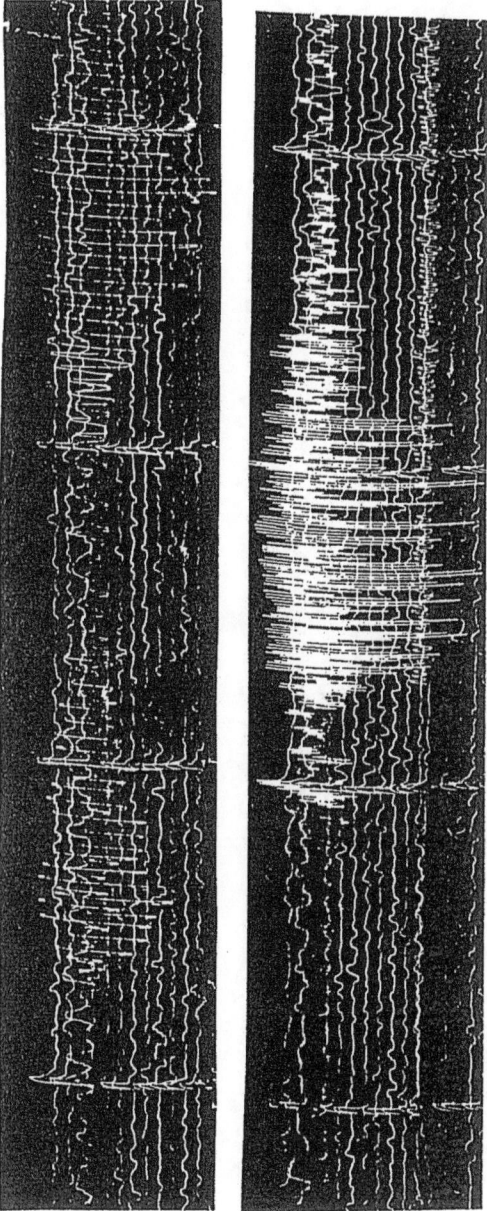

Figure 3.2 Wigwam recorded at Berkeley. The P-wavetrain, upper left, and the S-wavetrain, upper right, are atypically sinusoidal. The T-phase, lower section, arrived at an average velocity of about 1.75 km/sec, consistent with traveling most of the distance in the ocean and the last portion as a P-wave in the earth. Time deflections are at 60 sec. intervals.

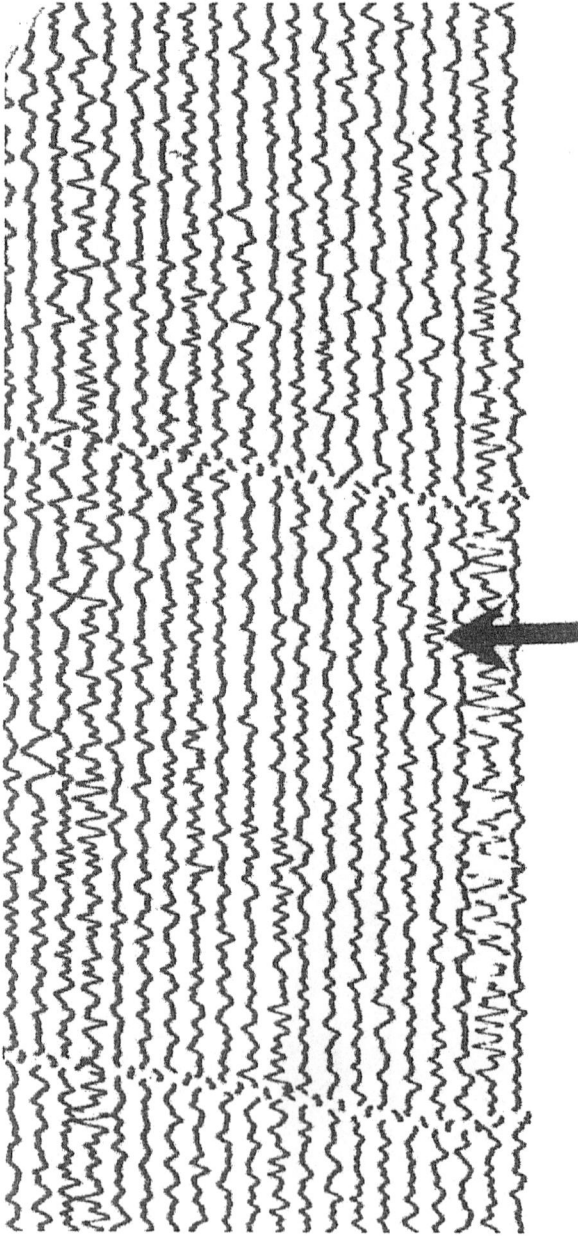

Figure 3.3 P-wave from Rainier as recorded at Fairbanks, Alaska. Time markers are at 60 second intervals. Note the small earthquake signal slightly less than 30 minutes after Rainier (second line from bottom).

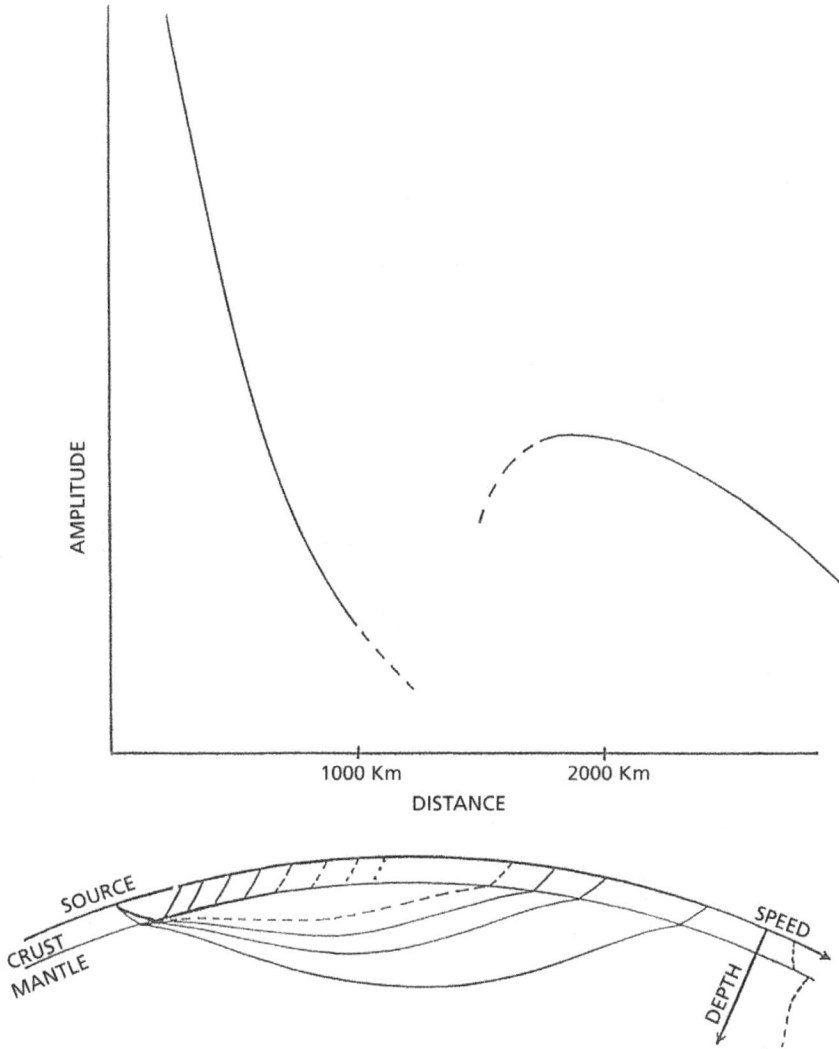

Figure 3.4 Amplitude and ray paths of P at intermediate distances. Waves guided along the base of the crust (Pn) are first arrivals between about 150 and 1000 kilometers. The amplitude of Pn falls off rapidly with distance. Beginning at about 1500 kilometers, energetic rays which penetrate the upper mantle emerge, but not necessarily as first arrivals.

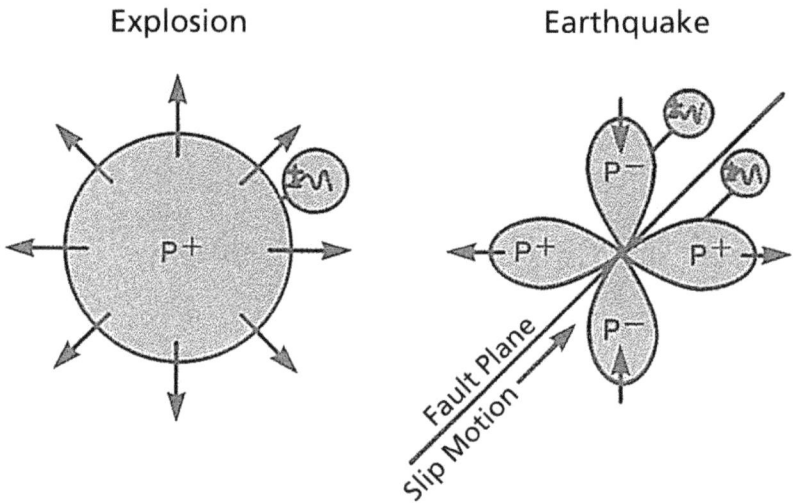

Figure 4.1. Right figures: Schematic representation of pattern of P-wave motion resulting from slippage along a fault line. P-waves have initial compressional motions in two quadrants, and initial rarefactional motions in the other two quadrants. Shear waves are also radiated perpendicular to the fault.
Left figure: Schematic representation of motion resulting from an explosion. P-waves have initial compressional motion in all directions. No shear waves are generated.

Figure 5.1 The Conference of Experts meeting in the Palais des Nations. United Nations officials at the head of the table; Eastern delegations to the left of the table, Western Delegations to the right.

Figure 5.2 Western Experts. At the table, left to right: William Penney, James Fisk, Robert Bacher, Yves Rocard, Ormand Solandt. Against the wall, right to left: Harold Brown, Jack Oliver, Normal Haskell, Perry Byerly, Unknown (State Dept.), Anthony Turkevitch, Carson Mark, Herbert Scoville, Carl Romney, Jack Morse, David Popper, George Olmstead, others unknown. Hans Bethe is seated directly in front of Jack Morse.

Figure 5.3 Seismologists (and a few others) at the Conference of Experts, 1958. Faces left to right: Frank Press, Carl Romney, Yu Riznichenko, Ivan Pasechnik, Pat Willmore, Doyle Northrup, (unknown), L.M. Brekhovskikh, K.E. Gubkin, M.A. Sadovsky, A. Zatopek, Norman Haskell, Jack Oliver.

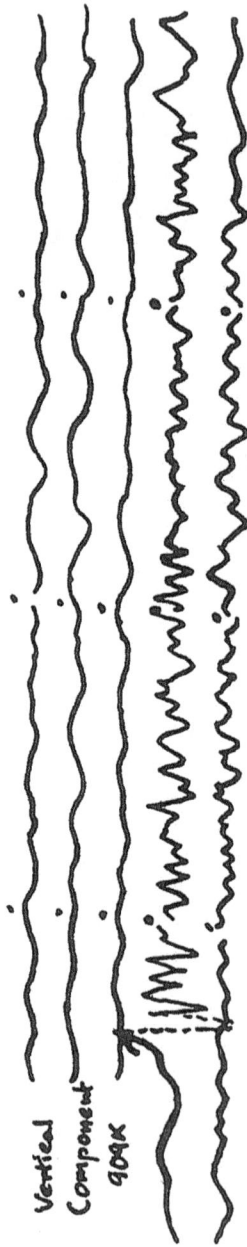

Figure 5.4 P-wave from an explosion of about 1 Mt as recorded by a Benioff seismograph at a distance of about 7,000 km. Time markers are at 10-second intervals. Ground motion magnified almost one million times (909 k).

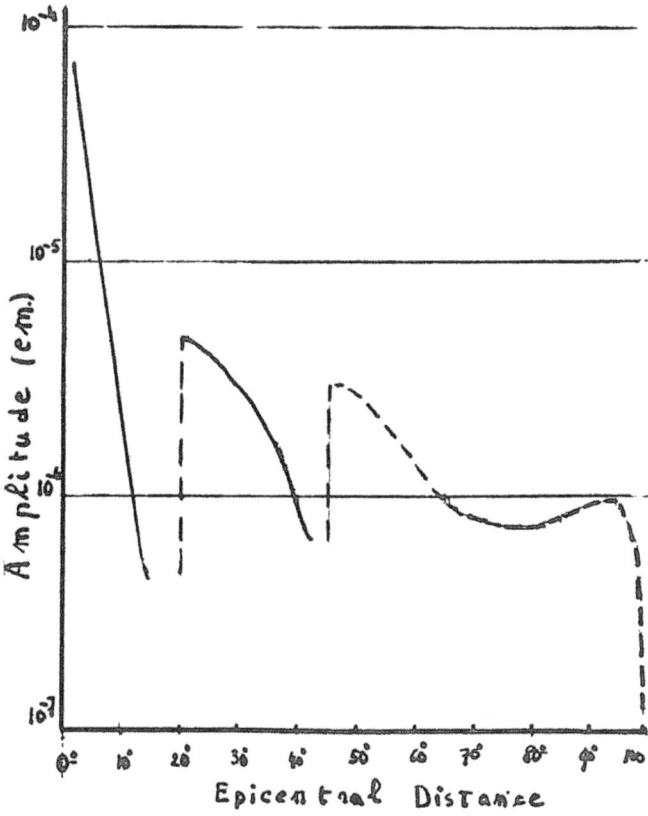

Figure 5.5 Amplitude of P-waves as a function of distance for an atmospheric explosion of about 1 Mt. Dashed portions were not well observed.

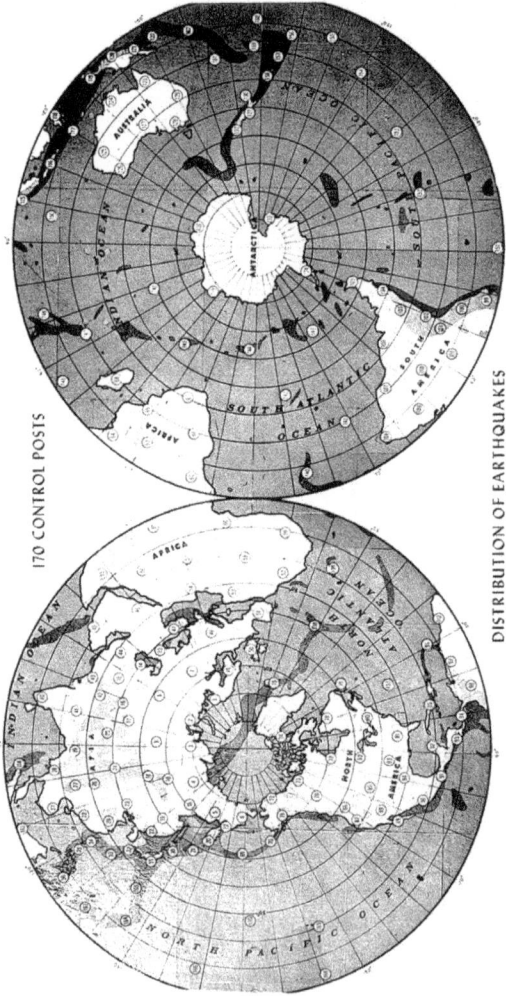

Figure 6.1 Location of control posts as proposed by Western Experts. More active earthquake zones are represented by heavy shading.

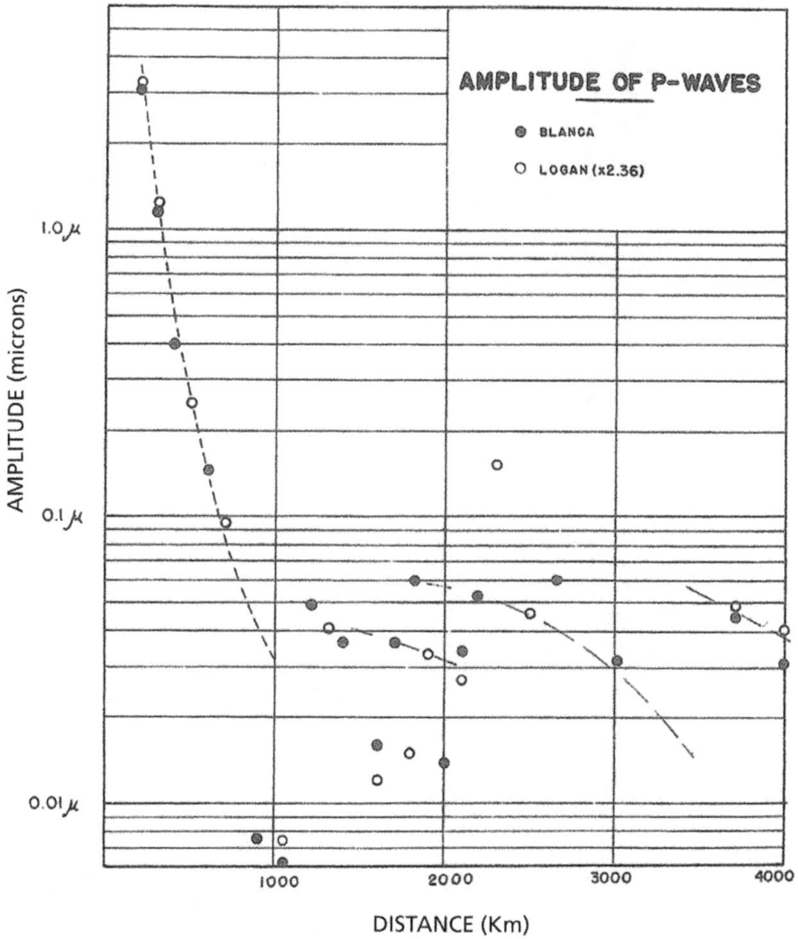

Figure 7.1 Amplitude of P waves from a 19-kt underground explosion as a function of distance. The dashed curve represents our interpretation in 1959.

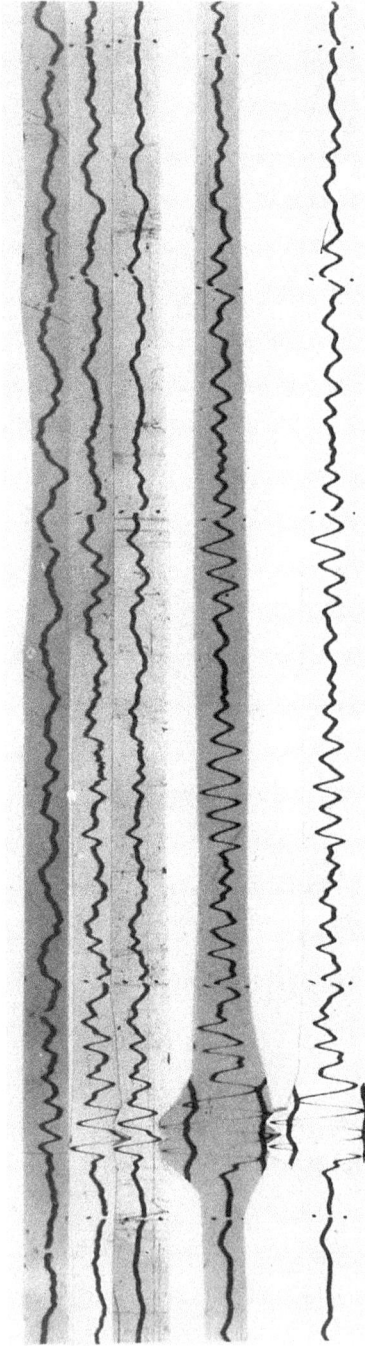

Figure 7.2 P-waves from Rainier, Logan, and Blanca as recorded at Ft. Sill, Oklahoma. Signal-to-noise ratio increases from top to bottom, revealing a precursor to the main P-pulse.

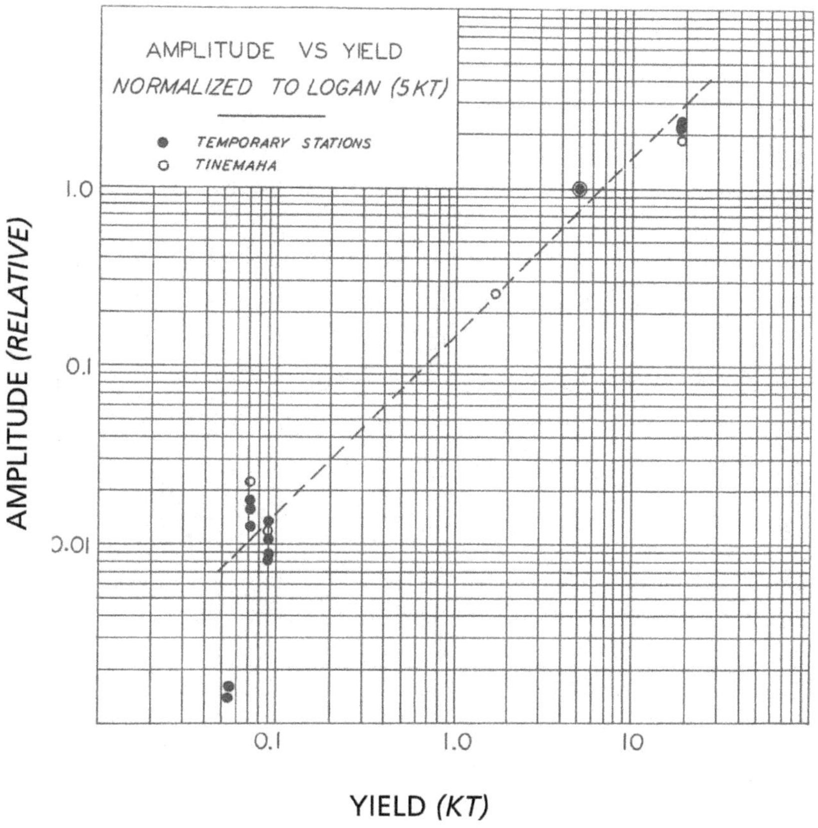

Figure 7.3 Amplitude of underground explosions as a function of yield. All measurements are relative to Logan (5 kt) amplitudes. Solid circles for Blanca (19 kt) are averages of university and temporary stations; solid circles for other shots are for individual station measurements.

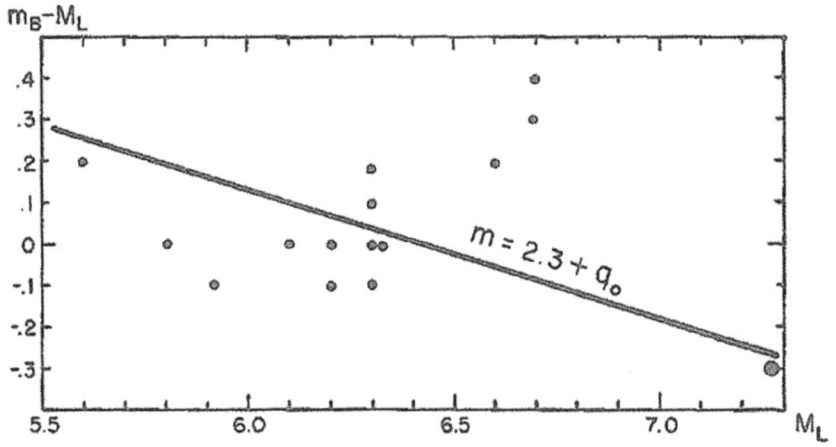

GUTENBERG – RICHTER   MAGNITUDE REVISION 1955

Figure 7.4 Figure from Gutenberg and Richter (1956 b). The line $M_L = m_b$ would be a straight horizontal line through zero on the left-hand axis.

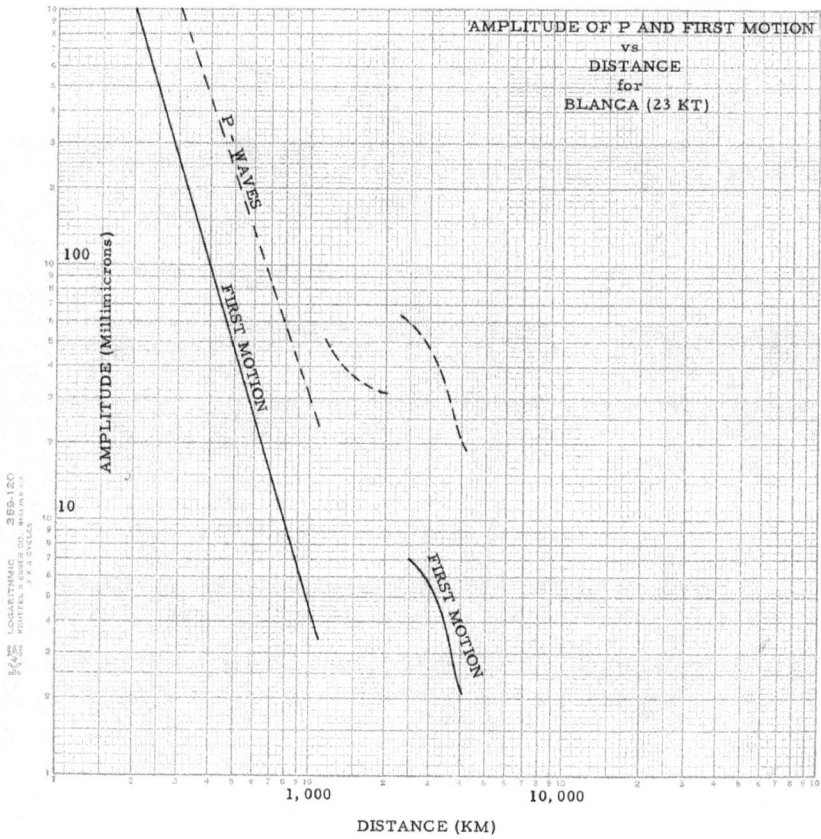

Figure 7.5 Amplitude vs distance for the main P-wave and for the first motion, normalized to *Blanca*.

Figure 9.1 TWG-II. At the table, left to right, Drs. Pnofsky, Fisk, and Romney. Seated behind, left to right, Drs. Press, Turkevitch, Bethe, and Tukey.

Figure 9.2 TWG-II. At the table, left to right, Drs. Pasechnik, Sadofsky, Fedorov, and Mr. V. Shustov. Standing, Drs. Riznichenko and Keilis-Borok. By the window to the right, Dr. Ustyumenko.

Figure 9.3 TWG-II Seated from left to right, Drs. Maddock and Penney at the table. Seated behind, second from left, Dr. J. W. Wright; most others are U.S. Conference Advisors.

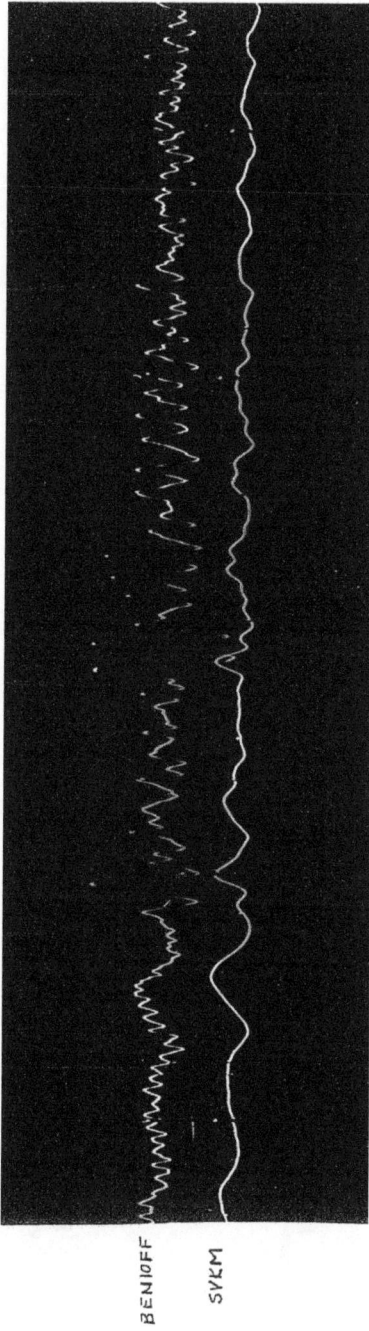

Figure 9.4 Comparison of a Benioff (top) and simulated SVK-M seismograms. Earthquake occurred at 00:31:43, 11 December 1958, near 5°S, 130° E

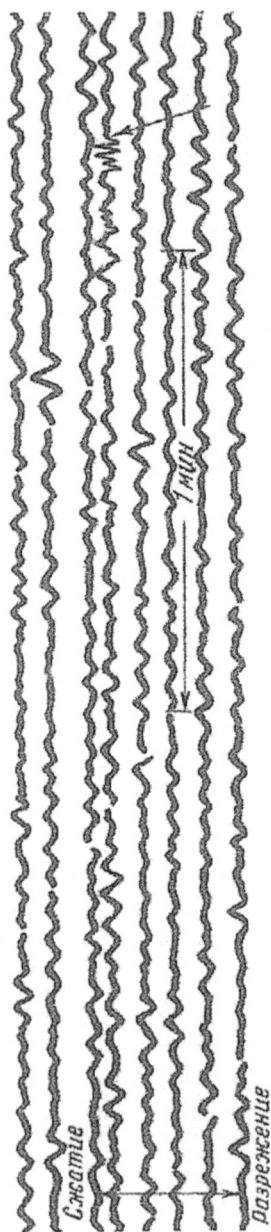

Фиг. 7. Запись взрыва Бланка на станции Тикси

$\Delta = 6890$ км, $V = 24$ тыс.

Figure 9.5 Tiksi recording of Blanca. Time increases from right to left. The P-wave signal marked by arrow, begins with apparent rarefactional motion. Length of 1-minute is indicated near center of recording.

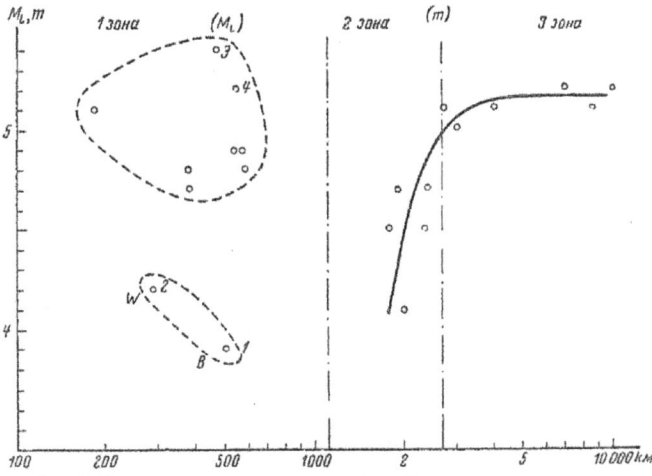

Фиг. 2. Зависимость измеренных магнитуд M_L и m от эпицентрального расстояния x для взрыва Бланка

1 — Баррет, 2 — Вуди, 3 — Маунт Гамилтон, 4 — Пало Алто

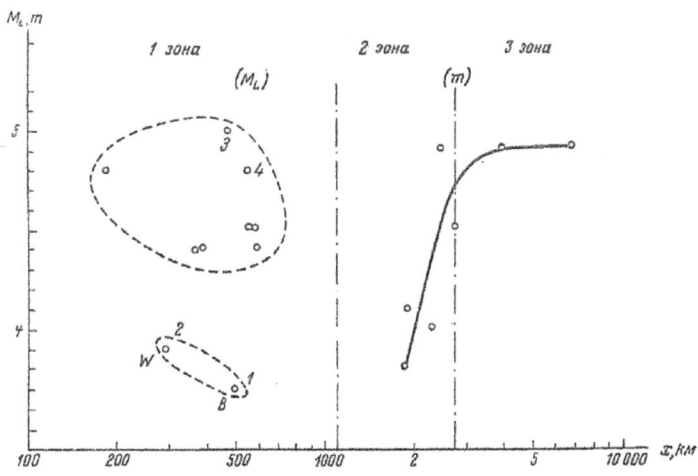

Фиг. 3. Зависимость измеренных магнитуд M_L и m от эпицентрального расстояния x для взрыва Логан

1 — Баррет, 2 — Вуди, 3 — Маунт Гамилтон, 4 — Пало Алто

Figure 9.6 Riznichenko's analysis of magnitude from Blanca (top) and Logan. Local magnitudes, M_L, have been "converted" to m_b using the Gutenberg and Richter formula. Magnitudes at Woody and Barrett are circled at lower left.

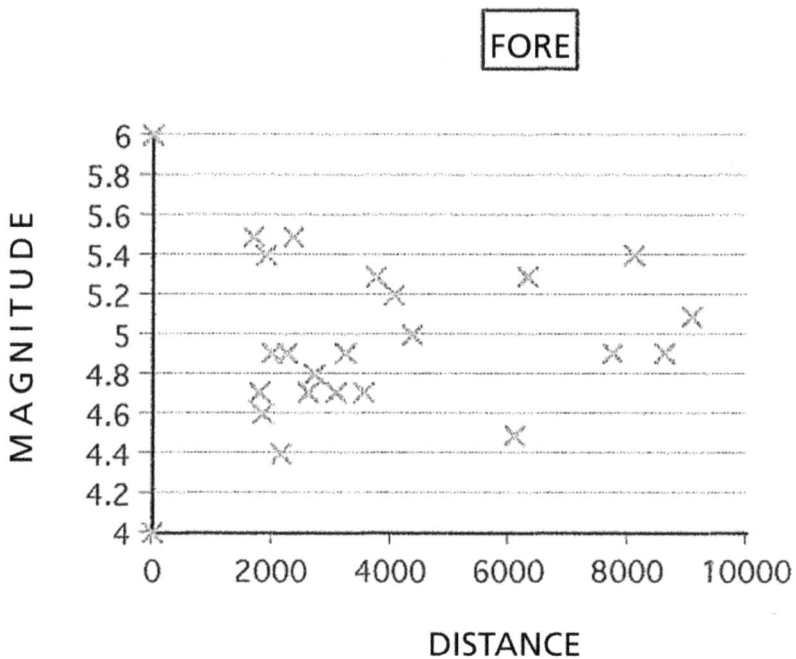

Figure 9.7 Magnitude (m_b) versus distance (km) for the underground nuclear explosion *Fore,* as measured at LRSM and WWSSN stations.

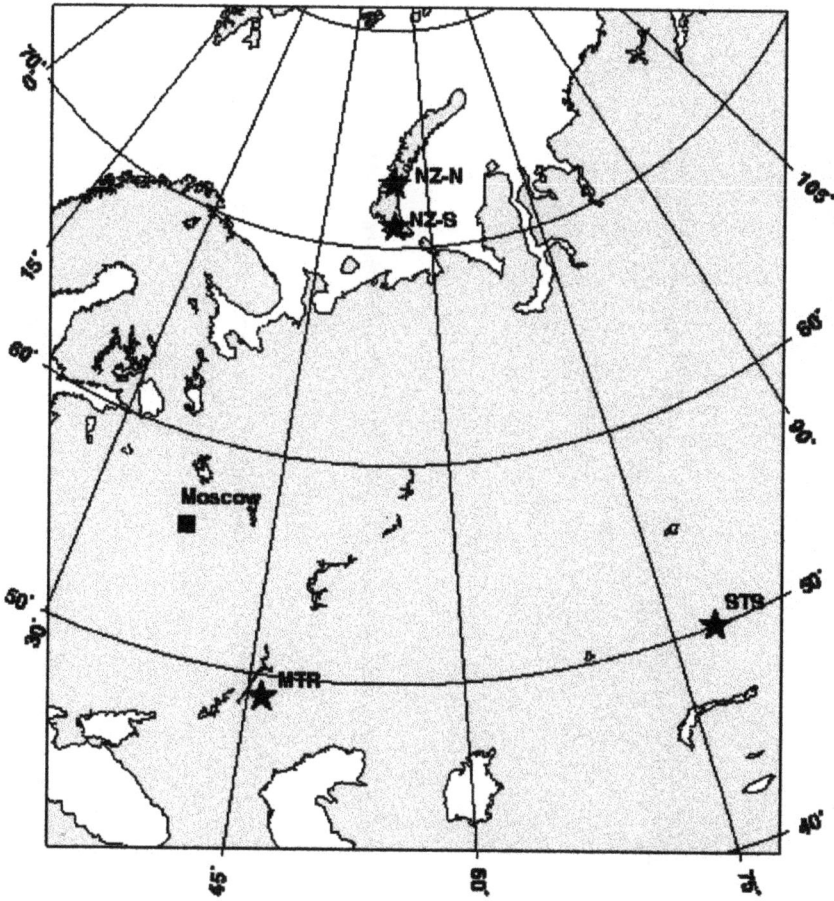

Figure 11.1 Soviet nuclear test sites during their 1961 test series, marked by stars. There was a northern and a southern testing are on Novaya Zemlya.

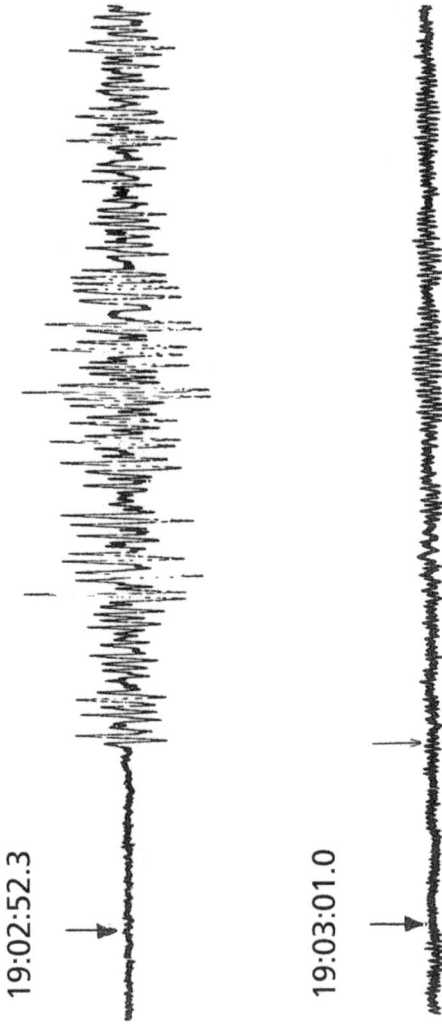

Figure 11.2 Variation in signal amplitude from Gnome. Upper trace: Jackson, Tennessee (1457 km), recording at magnification of 47k. Lower trace: Mina, Nevada (1465 km), recording at magnification of 712k.

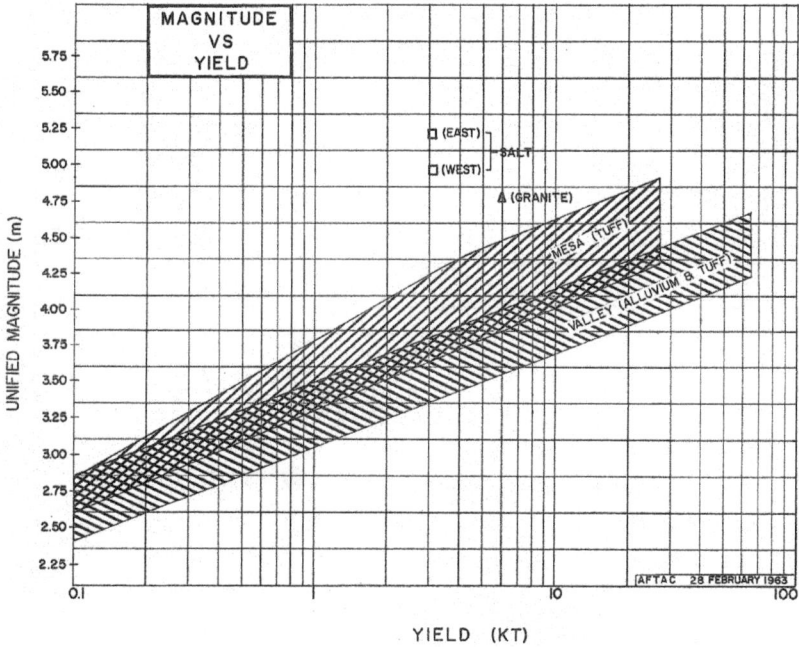

Figure 11.3 Magnitude (m_b) dependence on yield from NTS explosions and *Gnome*. Magnitudes of small shots have been scaled to large shots by the relative amplitudes of Pn, normalized to 500 km.

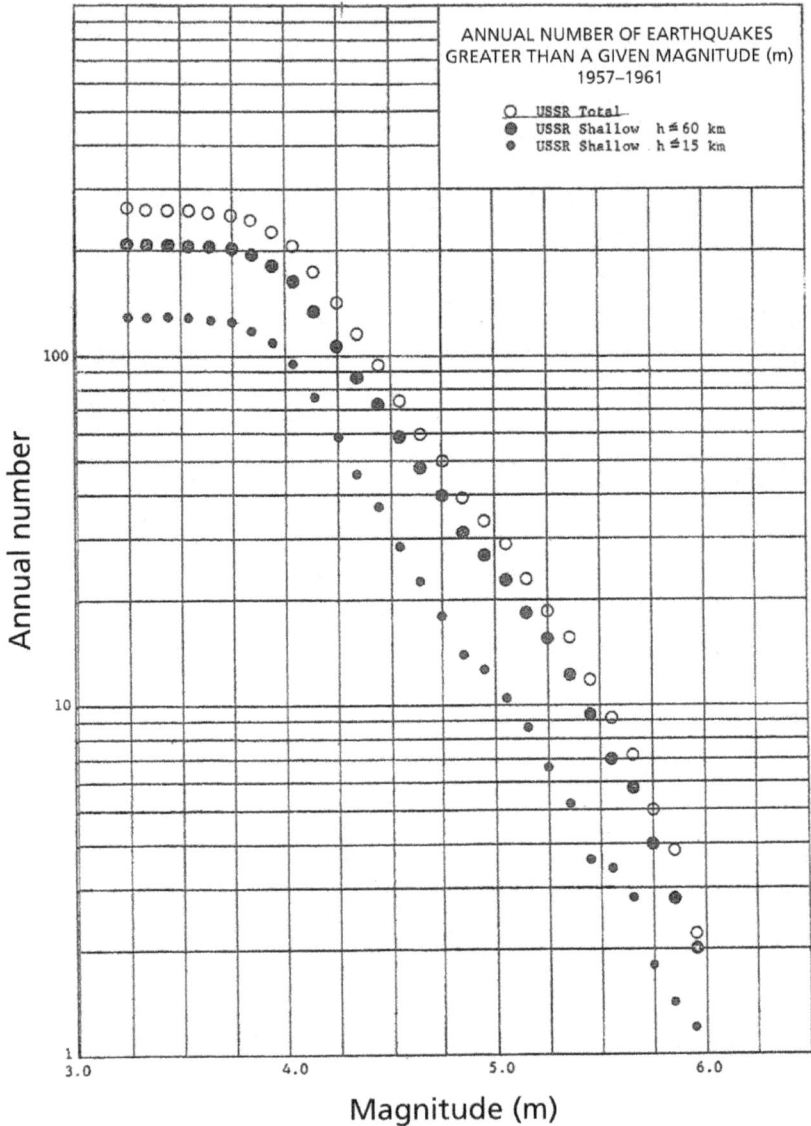

Figure 12.1 Annual number of earthquakes larger than a given magnitude in the USSR averaged for a five-year interval, 1957-1961. Only quakes deeper than 60 km were considered to be positively identified according to criteria then in use.

Figure 12.2 Earthquake of 14 November 1962 as recorded by a seismometer at the surface and at a depth of 8910 feet near Grapevine, Texas. Near-surface noise obscures the P-wave, and limits magnification to about 54 k. Greatly reduced noise at 8910 ft. depth permits signal to be clearly seen, and to operate the seismometer at a magnification of about one million at 1 Hz.

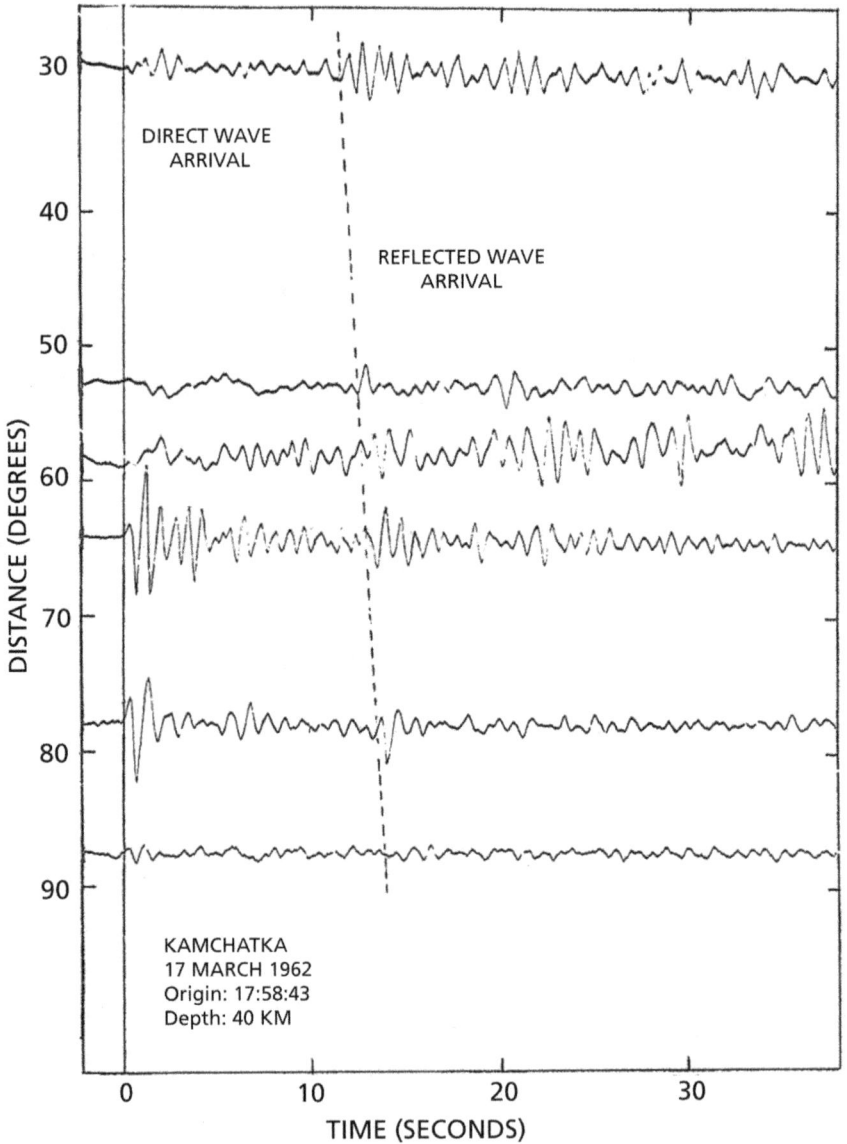

Figure 12.3 Seismograms, arranged in order of distance from a 40 km deep earthquake, show direct P-wave, and reflected wave, pP. Note that the time between direct and reflected wave increases with distance.

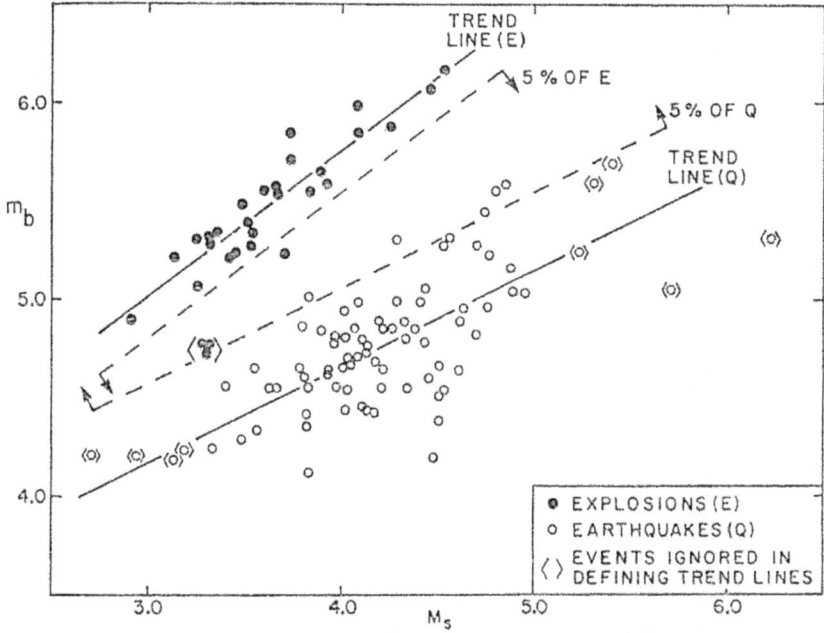

Figure 13.1 Discrimination between Eurasian earthquakes and underground explosions by the M_s:m_b criteria 1965-1970. The slope indicated for earthquakes is based on joint m_b and M_s data larger than about $m_b = 5$, the threshold for reliable detection of surface waves (Marshall and Basham, 1972).

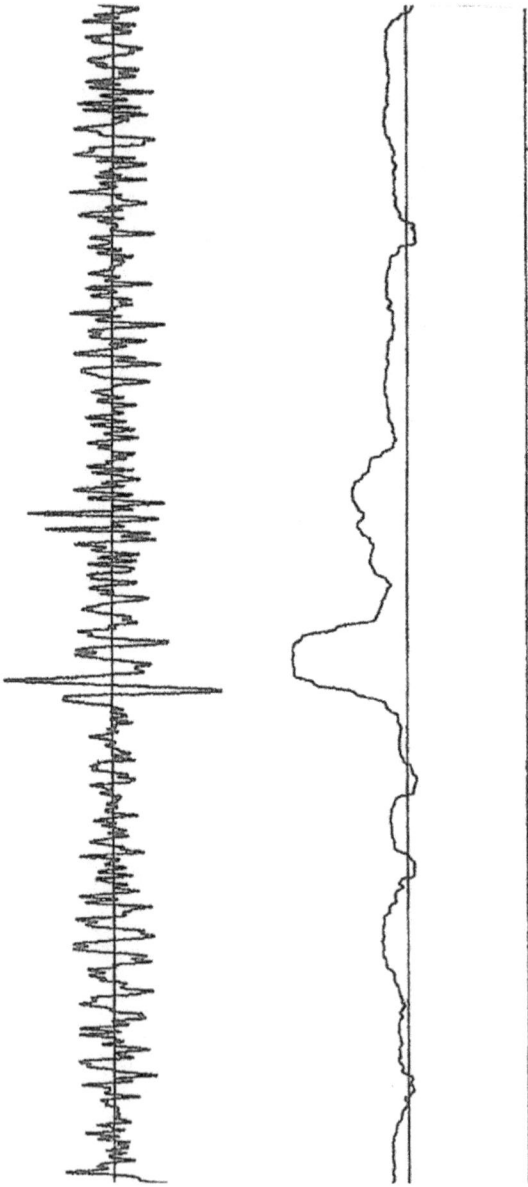

Figure 13.2 French Sahara explosion of 18 March 1967 as recorded at Tonto Forest Seismological Observatory. Upper trace: Signal on a single short-period sensor. Lower trace: Phased sums from two 10-element arrays of short-period sensors, multiplied together, integrated over a 3 second interval, and the square-root displayed.

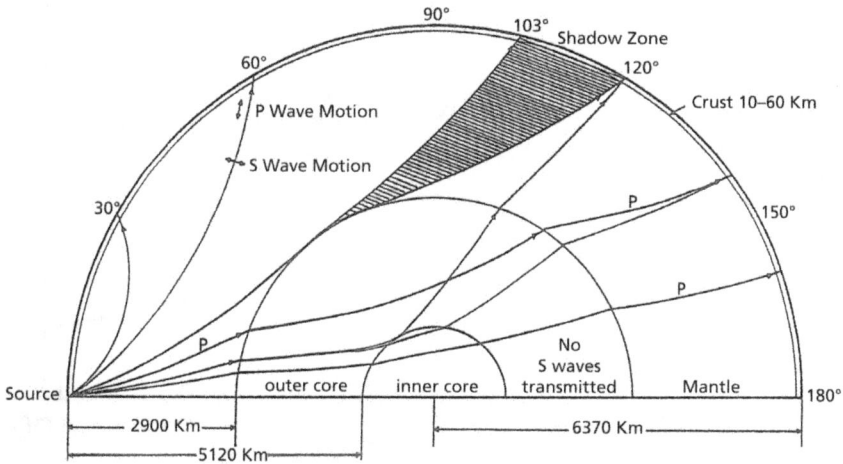

Figure A.1 Schematic cross section of the earth, showing major divisions and travel paths of seismic body waves. Seismologists measure distance around the earth in degrees; thus 90° is half-way around the earth, etc. Weak waves diffracted around the earth's core are recorded in the "shadow zone" beyond 103°. Shear waves cannot exist in a liquid, hence do not penetrate the core. A slight velocity increase in the inner core refracts (bends) P waves sharply upward to emerge at a distance of 120° and beyond. P and S waves are also reflected from the core and the underside of the crust (not shown).

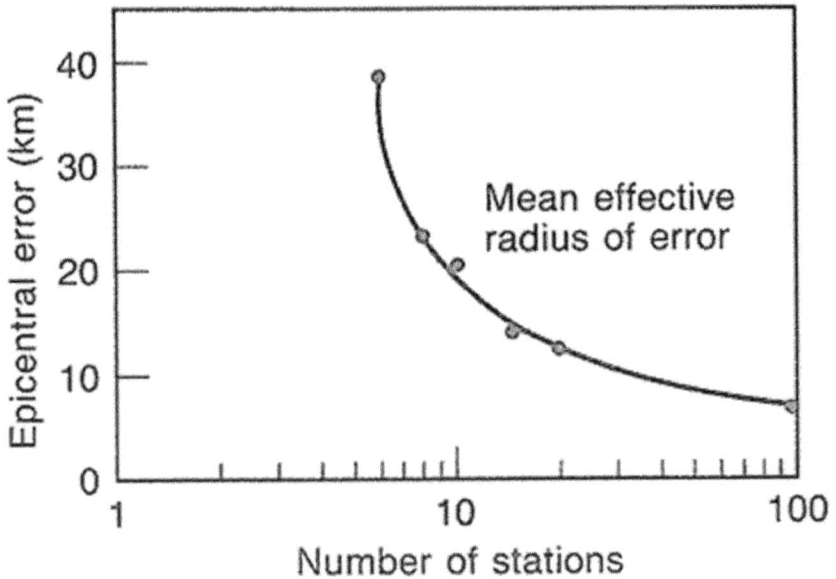

Figure D.1 Radius of a circle having the same area as the confidence ellipse about the epicenter. Calculated from subsets of stations that recorded NTS explosion *Bilby*.

Tables

Table 4.1

Yield	Detectable Distance
1 kt	Tens of miles to several thousand miles
10 kt	500 miles to 5,000 miles
100 kt	2,500 miles to any distance

Table 5.1 *Rainier*-Measured Amplitudes

Distance (km)	Amplitude (Millimicrons)	Period (Seconds)
280	70	0.3
820	8	0.5
920	2.5	0.6
1050	2.2	0.6
1560	Less than 1.5	No Signal*
3700	4.5	0.8

*Actually, a weak signal had been recorded but considered of dubious validity. Later it was validated by comparing its waveform to those from subsequent longer shots.

Table 5.2 **Number of Earthquakes Versus Magnitude**

m (Approx.)	N at Depth < 60 km
> 7.4	3
7.0 – 7.3	11
6.2 – 6.9	80
5.5 – 6.1	400
4.9 – 5.4	1,300
4.3 – 4.8	4,500
3.5 – 4.2	30,000 (?)

Table 7.1 **Key Ad Hoc Conclusions**

Kiloton Equivalent of Earthquakes	Annual Total of Earthquakes*	Percentage Identified	Annual Total of Unidentified Continental Earthquakes
>30 kt	170 – 700	95%	10 – 40
14 to 30 kt	270 – 1100	80%	50 – 200
7 to 14 kt	600 – 2300	25%	450 – 1700
3 to 7 kt	1100 – 4400	10%	1000 – 4000

*The ranges shown in these columns reflect the uncertainty in equating earthquake magnitudes to kiloton yield of underground nuclear explosions.

Table 8.1 **Estimated Annual Number of Unidentified Worldwide Continental Earthquakes**

	5 KT and Greater	10 KT and Greater	20 KT and Greater
Estimate Geneva Conference of Experts *August 1958*	20–100		
Estimate Geneva Network and Equipment on basis of Hardtack data *January 1959*	1500	400	60
Estimate Geneva Network with Improvements within the present state of technology on basis of Hardtack data *April 1959*	300	40	15

Table 8.2 **The Berkner Panel Budget Proposal**

	First Year	**Second Year**
Individual Research Projects (*15 project areas listed*)	$6.575 M	$10.520M
System Development (*Unattended stations, etc.*)	$4.250M	$7.480M
Nuclear and H.E. detonations	$12.000M	$12.000M

Table 9.1

Yield	**Annual Earthquakes**
1 kt	15,000/year
5 kt	5,000/year
20 kt	2,000/year

Table 9.2

	Magnitude in Working Paper of 5 Jan. 1959	Corrected Magnitude
Rainier (1.7 kt)	4.07 ± 0.4	4.7 ± 0.1
Logan (5 kt)	4.4 ± 0.4	4.95 ± 0.1
Blanca (19 kt)	4.8 ± 0.4	5.2 ± 0.1

Table 9.3 **Numbers of earthquakes equivalent to, or larger than explosions of given yield**

Yield (KT)	Tentative estimates 1959	Geneva estimates 1958	Working paper of 5 Jan. 1959
1	3,000	10,000	26,000
5	1,500	3,800	5,800
10	1,000	2,400	3,000
20	800	1,500	1,600

TABLE E.1 **Amplitude[a] of P waves versus distance**

BLANCA		LOGAN	
Distance km	Amplitude a/t mμ /sec	Distance km	Amplitude a/t mμ /sec
203.5	3,330	96.2	1,860.(P)
300.6	1,160		14,500.(Pn)
395.1	400	203.5	1,500
599.7	148	300.6	528
908.9	7.7	498.9	107
1036.0	5.2	714.5	41
1215.1	49	1036.0	3.1
1398.3	37	1111.5	. . .[b]
1610.1	16	1313.1	17.4
1707.0	37	1610.1	5.1
1842.0	60	1803.7	6.3
2011.2	14	1902.1	13.9
2111.3	34	2111.3	11.4
2208.8	53	2305.0	64
2665.3	60	2506.0	19.5
3017.4	32	3017.4	<10[c]
3308.9	<10[c]	3502.0	<10[c]
3717.5	˜44[d]	3717.5	˜21[d]
4020.6	30.5	4020.6	17.2

[a]P-wave amplitudes measured as half the maximum peak-to-trough displacement in the first few cycles of motion divided by period.
[b]P waves lost in noise from passing train.
[c]Signal not detected; noise estimated at 5 to 10 millimicrons.
[d]Estimated from College records.

TABLE E.2 **Magnitudes from torsion seismographs based on shear-wave amplitudes.**

Station	Distance (km)	Magnitude (M_L)				
		Blanca	*Logan*	*Rainier*	*Tamal.*	*Nep.*
Tinemaha	180.7	5.1	4.8	4.2	3.1	2.9
Woody	289.1	4.2	3.9			
Riverside	370.8	4.7	4.4			
Pasadena	382.2	4.8	4.4	4.0		
Mt. Hamilton	482.0	5.4	5.0	4.7		
Barrett	502.8	3.9	3.7			
Palo Alto	530.4	5.2	4.8	4.4		
Berkeley	540.5	4.9	4.5	4.1		
San Fran.	556.6	4.9	4.5	4.2		
Mineral	583.0	4.8	4.4	4.3		

Note: Blank spaces in the table mean signals were initially considered too weak to be accurately measured.

TABLE E.3 **Magnitudes from Benioff seismographs based on P-wave amplitudes.**

Distance (km)	Magnitude (m_b)	
	Blanca	*Logan*
1706.9	4.5	...
1803.7	...	3.8
1842.0	4.7	...
1902.1	...	4.1
2011.2	4.1	...
2111.3	4.5	4.0
2208.8	4.7	...
2305.0	...	4.9
2506.0	...	4.5
2665.3	5.1	...
3017.4	5.0	<4.5
3308.9	<4.5	...
3502.0	...	<4.5
4020.6	5.1	4.9

Glossary

Acoustic waves: The waves that transport sound energy through air or water.

Amplitude: The amount of motion of a wave in the earth or on a seismogram, measured from the mean (at rest) position to a maximum or minimum.

Array: An orderly arrangement of seismometers, placed at locations selected such that their combined outputs affect either signals or noise. As used here, arrays are designed to enhance the ratio of signal amplitude to noise amplitude.

Attenuation: Reduction in energy through absorption and other processes as seismic waves travel through the earth.

Azimuth: Direction; angle from north, measured clockwise.

Body waves: Either longitudinal (P) or transverse (S) seismic waves transmitted inside the earth.

Caustic: Region where seismic waves are focused.

Compressional motion: Wave motion that reduces the volume of an earth particle (squeezes) as it passes, as when the wave pushes the earth away from the source.

Coupling: Connection or linking of energy into seismic waves; the fraction of the total energy of an explosion that is transformed into seismic waves.

Damping: A force opposing vibration which acts to decrease the amplitudes of successive vibrations.

Diffraction: Deflection of waves into regions not directly exposed to the waves, by the edge of an obstacle or a caustic.

Epicenter: Location on the earth's surface directly above the focus of an earthquake.

Erg: A small basic unit of energy.

Event: A physical happening at a point in space and time. (See seismic event.)

Focus (of an earthquake): The location where is originates, including its depth.

Frequency (of a wave): The number of vibrations per second.

GMT: Greenwich mean time. Classical seismology is a global science, frequently using data from stations in different local time zones.

HE: High explosive (chemical rather than nuclear).

Horizontal component seismograph: A seismograph that is sensitive to horizontal motions of the earth: usually to motions in E-W and N-S directions.

Kilometer: 1/10,000 of the distance from earth's pole to the equator (0.621 mile).

Kiloton: 1,000 tons. Energy equivalent to 1,000 tons of TNT.

Long-period seismometer: A seismometer sensitive to waves with periods of oscillation of up to 10 seconds or more.

Longitudinal waves: Waves that travel in the earth, characterized by alternate compressional and rarefactional motions (like sound waves in air) and by motion-of-the-earth-in-line with the direction of travel of the waves; commonly called P-waves.

Love waves: Surface waves with motion in a transverse (perpendicular to the direction of travel), horizontal direction; named after famed mathematician A.E. H. Love.

Magnitude (of an earthquake): An arbitrary scale characterizing the size of an earthquake. Several scales have been devised. They are designed such that an increase by one magnitude unit, i.e., from magnitude 5 to 6, means that the seismic wave motions have increased by a factor of ten.

Megaton: 1,000,000 tons. Energy equivalent to one million tons of TNT.

Microseisms: Persistent feeble earth tremors due to natural causes such as winds or ocean waves. Also referred to as ambient or background noise.

Noise: Motions of the earth or on a seismogram that may interfere with the detection of a signal, e.g., microseisms or man-made disturbances.

P-waves: See longitudinal waves. (Originally from "Primary" or first arriving wave from an earthquake.)

Period: The time between two successive maximum (or minimum) motions of a wave or vibrating object. The inverse of frequency.

Pn: Longitudinal (or P) waves that propagate along the base of the earth's crust.

Rayleigh waves: Surface waves with motion in a vertical-radial plane. An earth particle moves in an elliptical pattern as Rayleigh waves pass, with the major axis vertical and motion at the top backward toward the source. Named after the famed physicist, R.J. Struitt (Lord Rayleigh).

Rarefactional motion: A wave motion that expands the volume of an earth particle as it passes, as when the wave pulls the earth back toward the source.

Refraction: Change in direction of waves caused by change in propagation velocity. The familiar glass lens exhibits this property as it bends light rays.

S-waves: See shear waves. (Originally from "Secondary" or second group of waves from an earthquake.)

Seismic area: An area in which earthquakes occur.

Seismic Event: A disturbance in the earth (usually an earthquake or explosion) that causes seismic waves to travel outward from a unique point and time of origin.

Seismogram: The visible record made by a seismograph showing ground motion as a function of time.

Seismograph: Instrument which senses and records seismic (earthquake) waves.

Seismometer: The motion-sensing component of a seismograph.

Shear waves: Waves that travel in the earth, characterized by motions of the earth perpendicular to the direction of travel of the wave train. Also called S-waves.

Short-period seismometer: A seismometer sensitive to waves having a period of one second or less.

Signal: Wave group of known type indicative of the occurrence of a seismic event.

SNR: Signal-to-noise ratio.

Sonic waves: Acoustic waves.

Surface waves: Waves that propagate along the earth's surface (similar to ocean waves); sometimes called L-waves.

Strata: Sections of rock formations that consist throughout of approximately the same kind of rock material.

Teleseism: A distant earthquake, beyond the zone where P and S waves in the crust and uppermost mantle are ordinarily observed, nominally beyond 2,000 km.

Transducer: The part of a device that converts motion into an electrical signal. A microphone is a familiar example.

Transverse waves: See shear waves.

Tuff: A type of rock formed from consolidated volcanic ash or particles.

Vertical component seismograph: A seismograph that is sensitive to up-and-down movements of the earth.

Yield: Energy released in a nuclear explosion, often expressed in terms of an equivalent weight of TNT.

Zipagram system: A method developed by AFTAC in the 1950's for transmitting seismic signals over leased telephone lines.

References

AFTAC, *A fifty year commemorative history of long range detection*, Hq. Air Force Technical Application Center, Patrick Air Force Base, Florida, 1997

Appleby, C.A. Jr., *Eisenhower and arms control, 1953-1961: A balance of risks*, 499 pages, Ph.D. dissertation, Johns Hopkins Univ., 1987.

Bailey, L. F., and C.F. Romney, *Seismic waves from the Nevada underground explosion of September 19, 1957*, unpublished, 1958.

Barth, K.H., *Detecting the Cold War: Seismology and nuclear weapon testing, 1945-1950*, 359 pages, Ph.D. dissertation, University of Minnesota, July 2000.

Båth, M., *Seismic records of explosions — especially nuclear explosions*, Part III, Swedish Defense Research Establishment, FOA Report, A270-A271, December, 192.

Benioff, H., "Earthquake seismographs and associated instruments," in *Advances in Geophysics*, Vol. 2, 200-275, Academic Press, H.E. Landsberg, Ed., 1955.

Berg, J.W., K.L. Cook, H.D. Narens, and W.M. Nolans, *Seismic observations of crustal structure in the eastern part of the Basin and Range Province*, Bull. Seis. Soc. Am., 50, 511-536, 1960.

Bethe, H., "Theory of seismic coupling," unpublished report presented to the Technical Working Group 2, Geneva, Switzerland, 7 December 1959.

Bird, K., and M. J. Sherwin, *American Prometheus: The Triumph and Tragedy of J. Robert Oppenheimer*, 721 pages, Alfred A. Knopf, New York, 2005.

Bolt, B.A., *Nuclear explosions and earthquakes: the parted veil*, W.H. Freeman and Company, San Francisco, 309 pages, 1976.

Bolt, B.A., *The revision of earthquake epicenters, focal depths and origin times using a high sped computer*, Roy. Ast. Soc., Geophys. Jour., 3, 473-440, 1960.

Brune, J., Espinosa, A., and J. Oliver, *Relative excitation of surface waves by earthquakes and underground explosions in the California-Nevada region*, Jour. Geophys. Res., 68, 3501-3513, 1963.

Brune, J.H. and P. W. Pomeroy, *Surface wave radiation patterns for underground nuclear explosions and small-magnitude earthquakes*, Jour. Geophys. Research, 58, 5005-5028, 1963.

Bullen, K.E., *Seismology in our atomic age*, Compt. Rendu. Assoc. Seism. et. Phys. Int. Terre, Strassbourg, 19-35, 1958.

Bullen, K.E., and T.N. Burke-Gaffney, *Diffracted seismic waves near the PKP caustic*, Geophys. J., 1, 9-17, 1958.

Burke-Gaffney, T.N. and K.E. Bullen, *Seismological and related aspects of the 1954 hydrogen bomb explosions*, Australian Jour. Phys., 10, No. 1, 130-136, 1957.

Byerly, P., *The seismic waves from the Port Chicago explosion*, Bull. Seis. Soc. Am., 36, 333-348, 1946.

Caloi, P., "About some phenomena preceding and following the seismic movements in the zone characterized by high seismicity," pgs. 44-68. in *Contributions in Geophysics* in honor of Beno Gutenberg, Pergamon Press,1958

Carder, D.S. and L. F. Bailey, *Seismic wave travel times from nuclear explosions*, Bull. Seis. Soc. Am., 48, 377-398, October, 1958.

Carder, D.S. and W.K. Cloud, *Surface motion from large underground explosions*, Jour. of Geophys. Res., 64, 1471-1497, October, 1959.

Carder, D.S., *Travel times from Central Pacific nuclear explosions and inferred mantle structure*, Bull. Seis. Sci. Am., 46, 2271-2294, 1964.

Carpenter, E.W., *An historical review of seismometer array development*, Proc. Inst. Electrical and Electronic Engineers, 53, 1816-1821, 1965.

Chung, D.H. and D.L. Bermreuter, *Regional relationships among earthquake magnitude scales*, Rev. Geophys. Space Sci., 19, 649-663, 1981.

Cole, A.C., and A. Goldberg, S.A. Tucker, and R.A. Winnacher, *The Department of Defense, documents on establishment and organization, 1944-1978*, Office of the Secretary of Defense Historical Office, 1978.

Dean, A.H., *Test ban and disarmament: the path of negotiation*, Oxford & CBH

Diment, W.H., S.W. Stewart, and J.C. Roller, *Crustal structures from the Nevada Test Site to Kingman, Arizona, from seismic and gravity observations*, Jour. Geophys. Res., 66, 201-214, 1961.

Doyle, H.A., *Seismic recordings of atomic explosions in Australia*, Nature, 180, 132-134, 1957.

Douze, E.J., *Signal and noise in deep wells*, Geophysics, 29, 721-732, 1964.

Engdahl, E.R., E.A. Flinn and C.F. Romney, *Seismic waves reflected from the earth's inner core*, Nature, 228, 552-555, 1970.

Gibowicz, S.J., *The relationship between teleseismic body-wave magnitude m and local magnitude ML from New Zealand earthquakes*, Bull. Seis. Soc. Am., 62, 1-11, February, 1972.

Gilpin, R., *American scientists and nuclear weapons policy*, 352 pages., Princeton University Press, 1962.

Goodman, Michael S., *Spying on the Nuclear Bear: Anglo-American Intelligence and the Soviet Bomb*, 295 pages, Stanford University Press, 2007.

Griggs, D.T., and E. Teller, *Deep underground test shots, UCRL-4659*, February, 1956

Griggs, D.T., and F. Press, *Probing the earth with nuclear explosions*, Jour. Geophys. Res., 66, 237, 258, January, 1961.

Gudzin, M.G. and J. H. Hamilton, *Wichita Mountains Seismological Observatory*, Geophysics, XXVI, 359-373, 1961.

Gutenberg, B., *Amplitudes of P, PP, and S and magnitude of shallow earthquakes*, Bull. Seis. Soc. Am., 35, 57-69, April, 1945.

Gutenberg, B., *Interpretation of records obtained from the New Mexico bomb test*, July 16, 1945, Bull. Seis. Soc. Am., 36, 327-330. October, 1946.

Gutenberg, B., *On the depth of the layer of relatively low wave velocity at a depth of about 80 kilometers*, Bull. Seis. Soc. Am., 38, 121-148, 1948.

Gutenberg, B., *Waves from blasts recorded in California*, Trans. Am. Geophys. Union, 33, 427-431, 1952.

Gutenberg, B., *Travel times of longitudinal waves from surface foci*, Proc. Nat'l Acad. Sci., 39, 849-853, 1953.

Gutenberg, B., *Low velocity layers in the earth's mantle*, Bull. Geol. Soc.Am., 65, 337-348.

Gutenberg, B., *The energy of earthquakes*, Quar. Jour. Geo. Soc. Lon. CXII, 1-14, August 28, 1956.

Gutenberg, B., *Caustics produced by waves through the earth's core*, Geophys. Jour., 15, 238-248, 1958.

Gutenberg, B., and C.F. Richter, *P'and the earth's core*, Mon. Not. R. Astron. Soc. Geophys. Suppl., 4, 363-372, 1938.

Gutenberg, B., and C.F. Richter, *Seismic waves from atomic bomb tests*, Trans. Am. Geophys. Union, 22, 776-778, 1946.

Gutenberg, B. and C.F. Richter, *Seismicity of the earth and associated phenomena*, 2nd. ed., 310 pp, Princeton Univ. Press, Princeton, New Jersey, 1954.

Gutenberg, B. and C. F. Richter, *Earthquake magnitude, intensity, energy and acceleration*, Bull. Seis. Soc. Am., Vol. 46 No. 3, 1956a.

Gutenberg, B., and C.F. Richter, *Magnitude and energy of earthquakes*, Annali di Geofisica, IX, 1-15, 1956b.

Guyton, J.W., "Systematic deviations of magnitudes from body waves at seismo-graph stations in the United States," in *Proceedings of VESIAC Conference on Seismic Event Magnitude Determination*, University of Michigan, 104 pgs, 1964.

Haskell, N.A., *An estimate of the maximum range of detectability of seismic signals*, Air force Surveys in Geophysics, No. 87, (AFCRC-TN-57-202), Air Force Cambridge Research Center, March 1957.

Heck, N.H., *Earthquakes, Facsimile of the edition of 1936*, 222 pp., Hafner Publishing Company, New York, 1965.

Hermann, R.B. and Nuttli, O.W., *Magnitude: the relation of ML to MbLg*, Bull. Seis. Soc. Am., 72, 389-397, April, 1982.

Herrin, E., *The resolution of seismic instruments used in treaty verification research*, Bull. Seis. Soc. Am., 72, S61-S67, 1983.

Herrin, E., and J. Taggart, *Regional variations in Pn velocity and their effect on the location of epicenters*, Bull. Seis. Soc. Am., 52, 1037-1046, 1962.

Holloway, D., *Stalin and the bomb, the Soviet Union and atomic energy, 1939-1956*. Yale University Press, 464 pages, 1994.

Howell, B.F., Jr. *Ground vibrations near explosions*, Bull. Seis. Soc. Am., 39, 285-310, 1949

Jacobson, H.K., and E.Stein, *Diplomats, scientists, and politicians*, 538 pp., Univ. of Mich. Press, Ann Arbor, 1966

Jeffreys, H., *On the Burton-on-Trent explosion*, Mon. Not. Roy. Aston. Soc., Geophys. Suppl., 5, 99-104, 1947.

Jordan, J., R. Black and C. Bates, *Patterns of maximum amplitudes of Pn and P waves over regional and continental areas*, Bull. Seis. Soc. Am., 55, 693-720. 1965.

Kim, W.Y. and P. G Richards, *North Korean Nuclear Test: seismic discrimination of low yield*, Eos, 88;11, 3 April 2007.

Kristianowsky, G.B., *A scientist at the White House*, 448 pp., Harvard Univ. Press, Cambridge, Ma., 1976

Lamb, H., *On the propagation of tremors over the surface of an elastic solid*, Philos.

Trans. Roy. Soc. London, Ser. A, 203, 1-42, 1904

Lapp, R;E., *Atoms and people, Harper and Brothers*, New York, 304 pp, 1956.

Latter, A.L., R. E. LeLevier, E. A. Martinelli, and W. G. McMillan, *A method of concealing underground nuclear explosions*, Jour. Geophys. Res., 66, 943-946, March, 1961.

Latter, A.L., E. A. Martinelli., and E. Teller, *A seismic scaling law for underground explosions*, The Rand Corporation, Santa Monica, Ca., January 1959.

Leet, L.D., Earth motion from the atomic bomb test, American Scientists, 60. 198-211, 1946

Lehman, Inge, *The travel times of the longitudinal waves of the Logan and Blanca explosions and their velocities in the upper mantle*, Bull. Seis. Soc. Am., Vol. 52, No. 3, 519-526, July, 1961.

Luksik, S., as reported in *Hearings before the Joint Committee on Atomic Energy*, October 28 and 28, 1971.

Mabon, D. W., D. S. Paterson and F.W. LaFantasie, *Arms control and disarmament*, Volume VII, U.S. Government Printing Office, 1995.

Macelwane, J.B. and F.W. Sohon, *Introduction to theoretical seismology*, Part I Geodynamics, John Wiley and Sons, Inc., 1936

Marshall, P., *On-site inspection for nuclear test ban verification*, Anali di Geofiva, XXXVII, 451-453, 1998.

Marshall, P. and P.W. Basham, *Discrimination between earthquakes and underground explosions employing an improved Ms scale*, Geophys. Jour. Roy. Astrol. Soc., 28, 431-458, 1972.

Melton, B.S., *Earthquake seismograph development: a modern history*, EOS, 62, No. 23, 505-510, May 1981 and 62, No. 24, 545-546, June 1981.

Murphy, J.R., *Types of seismic events and their source descriptions, in Monitoring a Comprehensive Test Ban Treaty*, Kluwer Academic Publishers, 1996.

Murphy, J.R., I.A. Kitor, B. W. Barker, and D.D. Sultanov, *Seismic source characteristics of Soviet peaceful nuclear explosions*, Pure and Appl. Geoph. 2077-2101, 158, 2001.

North, R.G., *Station magnitude bias — its determination, causes and effects*, MIT Lincoln Lab Technical Note, 1977-24, 1977.

Northrup, D. L. and D.H. Rock, *The detection of Joe 1, CIA, Studies in Intelligence*, Fall, 1966..

Oliver, J. and M. Ewing, *Seismic surface waves at Palisades from explosions in Nevada and the Marshall Islands*, Proc. Nat. Acad. Sci., 44, 780-785, August, 1958.

Oliver, J., and M. Ewing, *Short period oceanic and surface waves of the Rayleigh and first shear modes*, Trans. Am. Geoph. Union, 39, 482-485, 1958.

Panel on Seismic Improvement, *The need for fundamental research in seismology*, U.S. Dept. of State, 1959.

Pasechnik, I.P., *Characteristics of seismic waves in nuclear explosions and earthquakes*, Izdatal'stvo Nauka, 190 pgs, 1970.

Pasechnik, I.P., S.D. Kogan, D.D. Sultenov and V.I. Tsilbul'skiy, *The results of seismic observations made during underground nuclear and TNT blasts, in Seismic Effect of Underground Explosions*, Transactions of the Institute of Physics of the Earth im. O. Yu. Shmidta, No. 15 (182), Moscow, USSR Academy of Sciences, 108 Pgs, 1960.

Pomeroy, P. and J. Oliver, *Seismic waves from high altitude nuclear explosions,* Jour. Geophy. Res., 65, 3445-3457, 1960.

Ponton, J., S. Mohrer, C. Maag, R. Shepavek and J. Massie, *Operation Buster-Jangle 1951,* Defense Nuclear Agency, DNA 6023F, 1982.

Ramirez, J.E., *Tripartite and direction of approach of microseisms, in Symposium on Microseisms,* National Academy of Science-National Research Council, Publication 306, December 1955.

Richter, C.F., *An instrumental magnitude scale,* Bull. Seis. Soc. Am., 25, 1-32, 1935.

Richter, C.F., *New dimensions in seismology,* Science, 128, 175-182, 1958.

Robertson, H., "Azimuthal measurement by array processing (AZMAP)," *Geophysics: The leading edge of exploration,* 50-51, November 1992.

Romney, C.F., *Discussion of paper by Marcus Bath, in Symposium on Microseisms,* 66-73, National Academy of Sciences — National Research Council, Publication 306, December 1953

Romney, C.F., *Amplitudes of seismic body waves from underground nuclear explosions,* Jour. Geophys. Res., 64, 1489-1498, October, 1959.

Romney, C.F., *An investigation of the relationship between magnitude scales for small shocks,* in Proceedings of the VESIAC conference on seismic event magnitude determination, edited by VESIAC staff, 83-92 Inst. of Sci. and Tech., Univ. of Mich., Ann Arbor, May 1964.

Romney, C.F., *Operation BUSTER/JANGLE,* Project 7.2 (JANGLE), Project 7.5 (BUSTER), Beers and Heroy Participation, in: Seismic waves from A-bombs detonated over a land mass, by J. Allen Crocker, Headquarters U.S. Air Force, Office for Atomic energy, DCS/O, 15 March 1952.

Romney, C.F., B.G. Brooks, R. H. Mansfield, D. A. Carder, J. N. Jordan., and D. W. Gordon, *Travel times and amplitudes of principal body phases recorded from Gnome,* Bull. Seis. Soc. Am., 52, 1057-1074, December, 1962.

Romney, C. F. and W. Helterbran, "Progress and promise in the study of the earth using nuclear explosions," in *Engineering with nuclear explosions,* U.S. Atomic Energy Commission, TID-7695, 229-238, 1964.

Sakharov, A., *Memoirs,* Alfred A. Knopf, New York, 1990.

Seaborg, G.T., *Kennedy, Khrushchev, and the test ban,* Univ. California Press, 320 pp., 1981.

Sharpe, J.A., *The production of elastic waves by explosion pressures. I. Theory and empirical field observations,* Geophysics, VII, 144-154, 1942.

SIPRI (International Institute for Peace and Conflict Research), *Seismic methods for monitoring underground explosions,* 130 pp., 1968.

Snelling, R.W., *Technical history of the Geneva nuclear tests conference,* U.K. Public Record Office, DEPE 19/18, 1963.

Stevens, J.L. and S.M. Day, *The physical basis of mb:Ms and variable frequency magnitude methods for earthquake/explosion discrimination,* Jour. Geophys. Res. 90, 3009-3020, 1985.

Tatel, H.E. and M.E. Tuve, "Seismic exploration of a continental crust," in *Crust of the Earth,* Geol. Soc. Am. Spec. Paper 62, 35-50, 1955.

Thirlaway, H. J.S., *Forensic seismology, in The Vela program: A twenty five year review of basic research*, Ann. U. Kerr, Editor, Defense Advanced Research Projects Agency, 964 pp., 1985.

Thurber, C.A., H.R. Quinn and P.G. Richards, *Accurate locations of nuclear explosions in Balupan, Kazakhistan, 1997 to 1998*, Geophys. Res. Lett., 399-402, 1993.

U.S. Congress, *Hearings before the Joint Committee on Atomic Energy*, "Developments in the field of detection and identification of nuclear explosions (Project Vela)," July 25, 26, and 27, 1961.

U.S. State Department, "Geneva Conference on the Discontinuance of Nuclear Weapon Tests, History and Analysis of Negotiations," Department of State Publication 7258, 1961.

Vasiliev, A.P., "The initiatory stage of creation of the long range detection system in the USSR," presented at the Second International Symposium on the History of Atomic Physics, Laxenburg, Austria, October, 1999.

Veith, K.F. and Clawson, G.E., *Magnitude from short-period P wave data*, Bull. Seis. Soc. Am., 62, 435-452, April, 1972.

Voss, E.H., *Nuclear ambush, the test ban trap*, 612 pp. Henry Regnery Company, Chicago, 1963.

VSC, *Complexity of P waves from blasts and earthquakes*, Vela Seismological Center, Air Force Technical Applications Center, Technical Note VSC-9, 5 pp, 1964.

VSC, *The Waveform envelope off*, Vela Seismological Center, Air Force Technical Applications Center, Technical Note VSC-1, 10 pp, 1963.

Welch, M., *From the beginning*, Post-Monitor, Vol. 4, No. 3, Vol. 4, No. 4, Vol. 5, No. 1, Vol. 5, No. 3, AFTAC Alumni Association, Patrick Air Force Base, Florida, 1996-1997.

Werth, G.C., Herbst, R.F. and D.L. Springer, *Amplitudes of seismic arrivals from the M disconuity*, Jour. Geophys. Res., 67, 1587-1610, 1962.

Westhusing, K., *Project Gnome volunteer seismological teams*, Geophysics, 28, 20-45, 1963.

Wright, J.K. and Carpenter, E.W., *The generation of horizontally polarized shear waves by underground explosions*, Jour. Geophys. Res., 67, 1957-1963, May, 1962.

York, H.F., The advisors, Stanford University Press, 1989.

Ziegler, O.A. and D. Jacobson, *Spying without spies, origins of America's secret nuclear surveillance system*, Praeger, Westport, Connecticut, 242 pages, 1995

www.ingramcontent.com/pod-product-compliance
Lightning Source LLC
Chambersburg PA
CBHW030409100426
42812CB00028B/2891/J